Idaho
Total Eclipse Guide

Commemorative Official Keepsake Guidebook

2017 Total Eclipse State Guide Series

Aaron Linsdau

Sastrugi Press

Jackson Hole

Sastrugi Press / Published by arrangement with the author

Idaho Total Eclipse Guide: Commemorative Official Keepsake Guidebook

The author has made every effort to accurately describe the locations contained in this work. Travel to some locations in this book is hazardous. The publisher has no control over and does not assume any responsibility for author or third-party websites or their content describing these locations, how to travel there, nor how to do it safely. Refer to official forest and national park regulations.

Any person exploring these locations is personally responsible for checking local conditions prior to departure. You are responsible for your own actions and decisions. The information contained in this work is based solely on the author's research at the time of publication and may not be accurate in the future. Neither the publisher nor the author assumes any liability for anyone climbing, exploring, visiting, or traveling to the locations described in this work. Climbing is dangerous by its nature. Any person engaging in mountain climbing is responsible for learning the proper techniques. The reader assumes all risks and accepts full responsibility for injuries, including death.
Park maps are courtesy of the National Park Service.

Sastrugi Press
PO Box 1297, Jackson, WY 83001, United States
www.sastrugipress.com
Quantity sales: Special discounts are available on quantity purchases by corporations, associations, and others. For details, contact the publisher at the address above.

Library of Congress Catalog-in-Publication Data
Library of Congress Control Number: 2017905380
Linsdau, Aaron
Idaho Total Eclipse Guide / Aaron Linsdau- 1st United States edition
p. cm.
1. Nature 2. Astronomy 3. Travel 4. Photography
Summary: Learn everything you need to know about viewing, experiencing, and photographing the total eclipse in Idaho on August 21, 2017.

ISBN-13: 978-1-944986-06-3
ISBN-10: 1-944986-06-5

508.4—dc23

Printed in the United States of America
All photography, maps and artwork by the author, except as noted.

10 9 8 7 6 5 4 3 2 1

Contents

Introduction

Thank you for purchasing this book. It has everything you need to know about the total eclipse in Idaho on August 21, 2017.

A total eclipse passing across the United States is a rare event. The last US total eclipse was in 1979. It traveled over Washington, Oregon, Montana, and the corner of North Dakota.

The next total eclipse over the US will not be until April 8, 2024. It will pass over Texas, the Midwest, and on to Maine. After that, the next coast-to-coast total eclipse will be in 2045!

It's imperative to make travel plans today. You will be amazed at the number of people swarming to the total eclipse path. Some might say watching a partial versus a total eclipse is a similar experience. It's not.

This book is written for Idaho visitors and anyone else viewing the eclipse. You will find general planning, viewing, and photography information inside. Should you travel to the eclipse path in Idaho in mid-August, be prepared for an epic trip. The state estimates a half million visitors will converge on Idaho.

The Sturgis motorcycle rally will end one week before the eclipse. The event had a half million attendees in 2015. Bikers often drive to Idaho and Wyoming before and after the event. Chances are those bikers will travel to see the eclipse afterward.

All hotels in the Idaho communities of Idaho Falls, Driggs, Victor, and Swan Valley are sold out as of the writing of this guide. Finding lodging along the eclipse path will be a major challenge.

Resources will be stretched far beyond the normal limits. Think gas lines from the late 1970s. It's common for the shelves of Victor's grocery stores to quickly become short on bread, bananas, and similar staples during the summer. Be prepared with backup supplies.

Many smaller Idaho towns are far from any major city. Idaho roads are slow. Please obey posted speed limits within all forest and park areas. Be cautious about believing a map application's estimate of travel time in Idaho.

People in all communities along the path of the total eclipse plan to rent out their properties for this event. With a major celestial event in the summer of 2017, be assured that Idaho "hasn't seen anything yet."

Is this to say to avoid east Idaho or other areas during the eclipse? Not at all! This guidebook provides ideas for interesting, alternative, and memorable locations to see the eclipse. It will be too late to rush to a better spot once the eclipse begins. Law enforcement will be out to help drivers reconsider speeding.

Please be patient and careful. Large, beautiful, but dangerous animals make the forests in Idaho their home. Throughout the year, over one hundred animals are hit in Wyoming's Grand Teton alone. Moose, bear, and bison are difficult to see along the highways.

You should feel compelled to play hooky on August 21. Ask for the day off. Take your kids out of school. They'll likely be adults before the next chance to see a total eclipse. Create family memories that will last a lifetime. Sastrugi Press does not normally advocate skipping school or work. Make an exception because this is too big an event to miss.

Wherever you plan to be along the total eclipse path, leave early and remember your eclipse glasses. People from all around the planet will converge on Idaho. Be good to your fellow humans and be safe. We all want to enjoy this spectacular show.

Visit www.sastrugipress.com/ideclipse for the latest updates for this book.

Author Information

Polar explorer and motivational speaker Aaron Linsdau's first book, *Antarctic Tears*, is an emotional journey into the heart of Antarctica. This Jackson Hole native ate two sticks of butter every day to survive. Aaron coughed up blood early in the expedition and struggled through equipment failures. Despite the endless difficulties, he set a world record for surviving the longest solo expedition to the South Pole.

Aaron teaches audiences how the common person can achieve uncommon results. He uses stories to build grit, teach courage, and show how to maintain a positive attitude in the face of adversity. He hopes that you will have a comfortable and enjoyable time watching the total eclipse in Idaho.

All About Idaho

Overview of Idaho

Idaho, the Gem State, is full of wide-open country. With a population topping 1.6 million in 2014, Idaho ranks as one of the least populated states in the United States. Other than famous potatoes, many may not know about Idaho. That will soon change with the total eclipse passing over the state on August 21, 2017.

Idaho is dominated by mountains, farmland, and rivers. The largest city in the state is Boise with a population under a quarter million people. The next most populated city in Idaho is Nampa with a mere 86,000 people. Manufacturing, health care, tourism, and agriculture are the largest industries in the state.

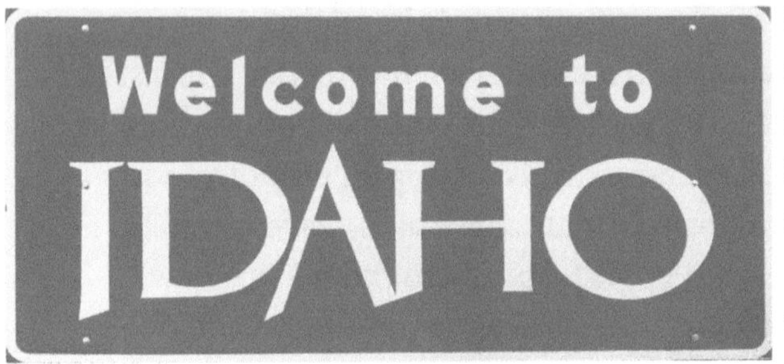

Fortunately for eclipse viewers, the northern part of the state generally receives more precipitation than southern Idaho with its warmer summer temperatures. Idaho winters can be long and particularly cold.

Yellowstone, the nation's first national park, is partly contained in Idaho. The state's panhandle even shares a forty-five-mile border with Canada. Grizzly bears, moose, and mountain lions are a few of the denizens of Idaho's wild backcountry.

Most people will visit the larger towns and cities in Idaho for the total eclipse. Thankfully, during the eclipse, the chance of snow and difficult weather is highly unlikely.

Hotels and Motels During the Eclipse

Driggs and Victor have seen a massive influx of people purchasing part-time or second homes in recent years. Many of the owners only stay for a few weeks while they enjoy winter activities in Idaho and Wyoming. The pressure from Jackson has significantly increased prices in Teton Valley as a consequence. Other towns in Idaho have not suffered from the same housing pressure.

Other cities in Idaho do not have the housing difficulties of Teton Valley. However, once word of the total eclipse over Idaho spread, rooms became scarce. Nearly every hotel in towns along the path of totality has been sold out for a year or more.

What does this mean for eclipse visitors? Lodging and room rentals in eclipse towns will be at a massive premium. Does that mean all hope is lost to find a place to stay? Not at all. But you will have to be creative. There will be few if any hotel rooms available in these eclipse cities by the time this book is printed. Accomodations in the cities and towns along the path of the eclipse have been sold out for months.

On March 29, 2017, the author searched on Hotels.com for rooms along the total eclipse path on the weekend of August 21 and found the following:

- Rexburg, ID, had rooms available at the Super 8 for $855 per night. They will likely be gone by the time you read this.
- Idaho: Few if no rooms were available in Idaho Falls, Pocatello, Blackfoot, Rexburg, Driggs, Victor, or Tetonia.
- Wyoming: No rooms are available in towns near Idaho. Jackson, Alpine, and Afton are all sold out.

Search for rooms farther away from the eclipse path. If you are willing to stay in Logan, UT, and drive two hours to Idaho Falls, rooms were available in March 2017. As the eclipse approaches, people will book rooms farther from the totality path. By midsummer, rooms in cities like Logan may be unavailable. The effect of this event will be felt across the country.

Tour buses routinely drive five hours from Salt Lake City to tour Jackson and Grand Teton and then drive back at night. Think regionally when looking for rooms. Be prepared to search far and wide during this major event. If a five-hour drive is manageable, your lodging options greatly expand, but it also increases your travel risk.

Yellowstone for East Idaho

Do not stay in towns or cities that require you to drive through Yellowstone to reach anywhere you want to be on the weekend of the eclipse. Traffic and unexpected animal jams can delay travel for hours. These jams are caused by tourists stopping to view popular animals. If wolves or bears appear, traffic will come to a complete standstill.

Do not believe any other map website when they show that travel through Yellowstone will only take an hour. Unless you are traveling late at night, add an extra hour to the travel estimate. The speed limit in Grand Teton and Yellowstone is heavily enforced.

Internet Rentals

To find rooms to stay in towns along the eclipse path, try a web service such as Airbnb.com. Note that many people rent out rooms or homes illegally, against zoning regulations. The towns of Driggs and Victor are starting to feel the crunch during summer months.

If Idaho towns fully enforces zoning laws, authorities may prevent your weekend home rental. Online home rentals during the eclipse will be a target for rental scams. People from out of the area steal photos and descriptions, then post the home for rent. You send your check or wire money to a "rental agent" then show up to find you have been scammed. If the deal sounds strange or too good to be true, run away.

Camping

If you can book a campsite, do it now. Do not wait. All areas in the national forests are first-come, first-served. Forest roads will be packed. Expect all areas to be swarming with people. Show up early to stake out your spot. Consider staying farther away and driving early on August 21.

Please respect private land too. Idaho folks don't take kindly to people overrunning their property without permission. In a big state with only 1.6 million residents, people are very protective, but they're friendly, too. You never know what you might be able to arrange with a smile and a bit of money.

This all said, there are plenty of camping opportunities throughout Idaho. You don't have to sleep exactly on the eclipse path. If you're ready to rough it, there are national forest camping options.

Government agencies have been meeting since 2016 to talk about how to manage the influx of people. Please note that every possible government agency will be working full time to enforce the various rules and regulations.

National Parks

Chances are you will not find a camping site in the national park. To watch the eclipse from Craters of the Moon National Monument and Preserve, you do not have to sleep in it. You just need to drive there in the morning.

Park law enforcement will be present on the eclipse weekend. Hundreds of thousands of people are expected in the region. Parking will overflow. It will make parking lots and lines on Black Friday at the mall look uncrowded. For an event of this magnitude, you'll need to find your parking space early.

The first sentence of the national parks mission statement is:

> *"The National Park Service preserves unimpaired the natural and cultural resources and values of the national park system for the enjoyment, education, and inspiration of this and future generations."*

Roadside camping (sleeping in your car) is not allowed in the park. Park facilities are only designed to handle so many people per day. Water, trash collection, and toilets can only withstand so much. If you notice trash on the ground, take a moment to throw it away. Protect your national park and help out. Rangers are diligent and hardworking but they can only do so much to manage the expected crowds.

National Forests

There are plenty of national forests in Idaho. They all have camping opportunities. The forest service manages undeveloped and primitive campsites. Be sure to check for any fire restrictions. Check with individual agencies for last-minute information and regulations. The Forest Service requires proper food storage. Plan to purchase food and water before choosing your campsite. Below is a list of national forests along or near the total eclipse path:

Boise NF:	www.fs.usda.gov/boise
Caribou-Targhee NF:	www.fs.usda.gov/ctnf
Payette NF:	www.fs.usda.gov/payette
Salmon-Challis NF:	www.fs.usda.gov/scnf
Sawtooth NF:	www.fs.usda.gov/sawtooth

Forest service roads abound in Idaho. Maps for forests are available at local visitor centers and bookstores. This book's website has digital copies of some forest maps.

Printed national forest maps are large and detailed. Viewing digital maps on your smartphone or pad is difficult. If you plan to camp in the forest, a real paper map is a wise investment.

Camping in federal wilderness areas is also allowed. Those areas afford the ultimate backcountry experience. However, be aware that no vehicle travel is allowed in the specially designated areas. This ban includes: vehicles, bikes, hang gliders, and drones. You can travel only on foot or with pack animals.

Sleep in Your Car

Countless RVs, campers, trucks, cars, and motorcycles will flood Idaho. Sleeping in your car with friends is tolerable. Doing so with unadventurous spouses or children is another matter.

Do not be caught along the path of the total eclipse without some sort of plan, especially in east Idaho. The whole area brims with people on a normal summer day. August 21 will be anything but normal.

Useful Local Webcams

Local webcams are handy to make last-minute travel decisions. The webcams are sensitive enough to show headlights at night. Use them to determine if there are issues before traveling out. Commuter traffic between Jackson, Wyoming and Victor is heavy.

The smartphone application Wunderground is useful to check on the webcams in one place. All the webcams are listed near the bottom of the app window.

Weather

It's all about the weather during the eclipse. Nothing else will matter if the sky is cloudy. You can be nearly anywhere in Idaho and catch a view of the sky when traffic comes to a standstill. But if there's a cloud cover forecast, seriously reconsider your viewing location.

Travel early wherever you plan to go. Attempting to change locations an hour before the eclipse due to weather will likely cause you to miss the event. Idaho roads are narrow and slow. The mountain passes are steep, and vehicles back up routinely.

Modern Forecasts

Use a smartphone application to check the up-to-date weather. Wunderground is a good application and has relatively reliable forecasts for the region. The hourly forecast for the same day has been rather accurate for the last two years. The below discussion refers to features found in the Wunderground app. However, any application with detailed weather views will improve your eclipse forecasting skills.

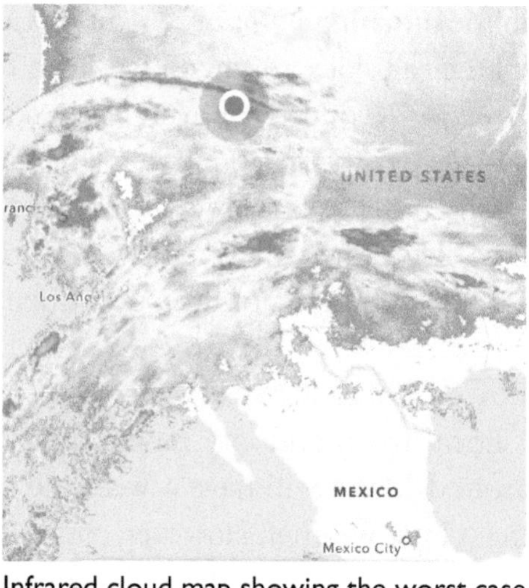

Cloud Cover Forecast

The most useful forecast view is the visible and infrared cloud-coverage map. Avoid downloading this app the night before and trying to learn how to read it. Practice reading them at home. It's imperative to understand how to interpret the maps early.

Infrared cloud map showing the worst case eclipse cloud cover.
Courtesy of National Weather Service.

All cloud cover, night or day, will appear on an infrared map. Warm, low-altitude clouds are shown in white and gray. High-altitude cold clouds are displayed in shades of green, yellow, red, and purple. Anything other than a clear map spells eclipse-viewing problems.

To improve your weather guess, use the animated viewer of the cloud cover. It will give you a sense clouds motion. You can discern whether clouds or rain are moving toward, away from, or circulating around your location.

Normal Idaho Weather Pattern

Due to the direction of the jet stream, most weather travels across Oregon and Montana, then into Idaho. On occasion, weather can approach from any direction. Due to the nature of the Rocky Mountains and the Sawtooth Range, weather in east Idaho can be unpredictable.

The common weather pattern in August is clear in the morning and cloudy in the afternoon in the western half of the state. As the day warms up, clouds develop over Idaho and begin flowing east. These clouds flow into Wyoming in the afternoon.

Cities east of any Idaho mountain ranges tend to have clear skies in the morning. Do not rely on this action though. If anything other than clear skies are predicted, plan to drive to other parts of Idaho, Oregon, Wyoming, or even Nebraska.

Consider that slow-moving clouds can obscure the sun for far longer than the two-minute duration of the totality. The time of totality is so short that you do not want to risk it. Missing it due to a single cloud will be a major disappointment.

Local Eclipse Weather Forecasts

Local town and city newspapers, radio, and television stations around Idaho will have a weekend edition with articles discussing the eclipse weather. However, conditions change unpredictably in the Rocky Mountains. A three-day forecast may be completely incorrect.

The website www.mountainweather.com will feature a paid section specifically focused on eclipse weather during the summer of 2017. Jim Woodmencey runs the site and is a Jackson, WY weather forecaster. He is regularly featured on local radio stations. Start watching the forecasts days in advance.

Forest Fires

For the past several years, forest fires have been common in the western United States. The summer of 2017 is likely to be no different. Forests in Idaho had major fires during most of the 2016 summer. Chances are there will be fires in the region in the summer of 2017. For fire updates check https://inciweb.nwcg.gov.

For the best eclipse viewing experience, you need to have as clear a sky as possible. The subtleties of the sun's corona will be obscured by fog, clouds, or smoke. If you think the view of the sky is going to be blocked, don't wait until the last minute to move to a clearer location.

Road Closures Due to Fires

Highways connecting various Idaho towns were closed for several days during the summer of 2016. The Pioneer Fire caused the closure of Idaho 21 near Idaho City in July and August of 2016.

In 2016, the Tepee Springs Fire prompted the Forest Service to close access to part of the Salmon River. All Nez-Perce-Clearwater National Forest land closed. Portions of the Payette National Forest closed.

All those fires were near or on the total eclipse path. If 2016 was any indication what the summer of 2017 will be like, count on fire that will affect eclipse viewing.

The above major events and smaller fires in the region completely blocked the view of the Tetons and Wind River Mountains in Wyoming for over a week. More than once, the noon sun was invisible. An eclipse at that time would have been obscured. The most accurate website for fires is:

inciweb.nwcg.gov

Check the Idaho road report for updated closure information:

hb.511.idaho.gov

It's imperative to plan for fires and their effects. Watch the weather reports. If strong winds and lightning storms are forecast, prepare to

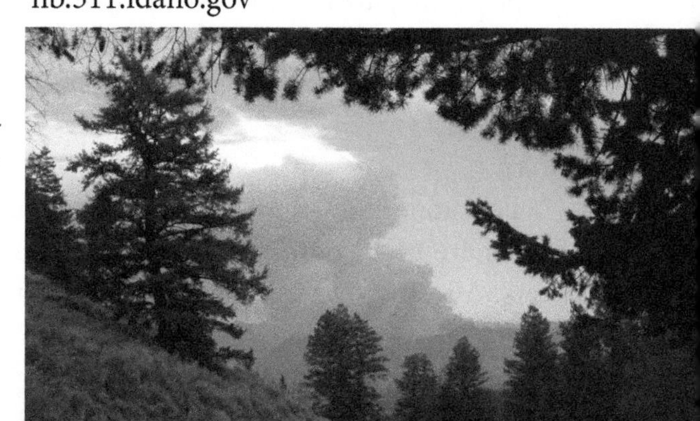

reconsider your viewing location. If conditions are poor, you and thousands of other vehicles will be trapped in slow-moving traffic.

If you believe it's necessary to leave a town to watch the eclipse, do so the night before or extremely early in the morning. RVs are common, and trains of them crawl over Togwotee and Teton passes.

Idaho Information

Cellular Phones

Cellular "cell" phone service in remote Idaho locations is spotty at best. Most of the time there is good coverage along the main highways and interstates. However, even along major thoroughfares, there is little or no coverage.

It's possible to find zones where text messages will send when phone calls are impossible. If you cannot make a phone call, the chance of having data coverage for web surfing or e-mail is low.

Please look up any information or communicate what you need before departing from the main roads around Idaho. Bureau of Land Management (BLM) areas sometimes have coverage. Planned to be self-contained. Treat like your cell phone like it won't connect.

You may find yourself out of cell service. With a large number of cell users in a concentrated area, coverage and data speed may collapse as well. Search on the phrase "cell phone coverage breathing".

National Park and Forest Safety

Nearly all regions of Idaho are full of wild animals. Although they are beautiful, the animals are dangerous. They can easily injure or kill people, as they are far more powerful than humans. Do not try to feed any wild animals, including squirrels, as they can carry the plague. These suggestions apply to all public lands in Idaho.

Bear

Yellowstone and Grand Teton National Parks and the surrounding forests are bear country. They are home to both grizzly and black bears. Always travel in groups of three or more. If a bear hears you, it will usually vacate the area. Bear charges are often caused by unexpected and surprise encounters.

Noise is the best defense to avoid surprising bears. Regularly clap, make noise, and talk loudly. The *Grand Teton Bear Safety* brochure states, "Bear bells are not sufficient. The use of portable audio devices is strongly discouraged."

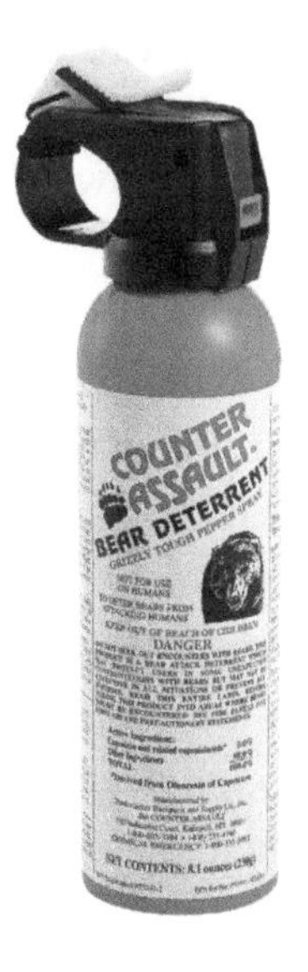

Always have bear spray while hiking. Carry it in a holster or in a cargo pocket. You need to practice being able to discharge the spray within three seconds. Carrying it in a backpack is completely ineffective.

Current park regulations require people to stay one hundred yards (300 feet) away from all bears and wolves. They are exciting to see but need their space. Refer to current park regulations for more safety information.

Moose

This member of the deer family is extremely defensive when they are with their young. If you see a moose calf, leave the area. If a moose approaches, back away. Put something between you and the moose. Unlike bears, it is okay to run from moose. Stay at least 25 yards away from moose.

Bison

Bison may appear tame and docile, but they are powerful wild animals. These 1,400-pound herbivores can run three times faster than humans. Bison do not like being crowded or harassed. Stay at least twenty-five yards (75 feet) away to avoid injury or death. Do not walk any closer to capture them with your camera. Search Google for "2015 bison gores girl" to see what can happen when people get too close.

Mountain Lion

If you encounter a mountain lion, do not run. Keep calm, back away slowly, and maintain eye contact. Do all you can to appear larger. Stand upright, raise your arms, or hoist your jacket. Never bend over or crouch down. If attacked, fight back.

Altitude Sickness

Anyone visiting the Rockies from a lower elevation might develop symptoms of altitude sickness. The symptoms can run from mild to unpleasant (or even dangerous). At Drigg's elevation of 6,109 feet, there is only 81% of the oxygen available at sea level. The lack of oxygen combined with low humidity can cause people to experience altitude sickness.

The symptoms of altitude sickness can strike anyone regardless of fitness level. They include:

- Headaches
- Nausea or vomiting
- Loss of appetite
- Insomnia
- Dizziness
- Fatigue
- Loss of energy

Avoiding Altitude Sickness

1. Hydrate

The number one suggestion to avoid altitude sickness is to drink plenty of water. Consume more than you would at home, especially if you live in a humid environment.

2. Acclimation time

Give your body time to acclimate. Mountaineers who climb high peaks ascend slowly. Give yourself a few days to adjust.

3. Sun exposure

The sun is much more intense in the mountains than at sea level. Wear sunglasses and liberally apply sunscreen to avoid sunburns.

4. Eat well

Keep your energy up. Appetite loss is common at higher altitudes Maintain your normal eating schedule.

5. Prepare for temperature changes

Temperatures will drop rapidly once the sun sets in the mountains. Bring appropriate clothing.

6. Talk with your doctor

If the altitude bothers you or you have experienced previous altitude problems, talk with your doctor before arriving. Seek professional medical attention if you develop serious symptoms.

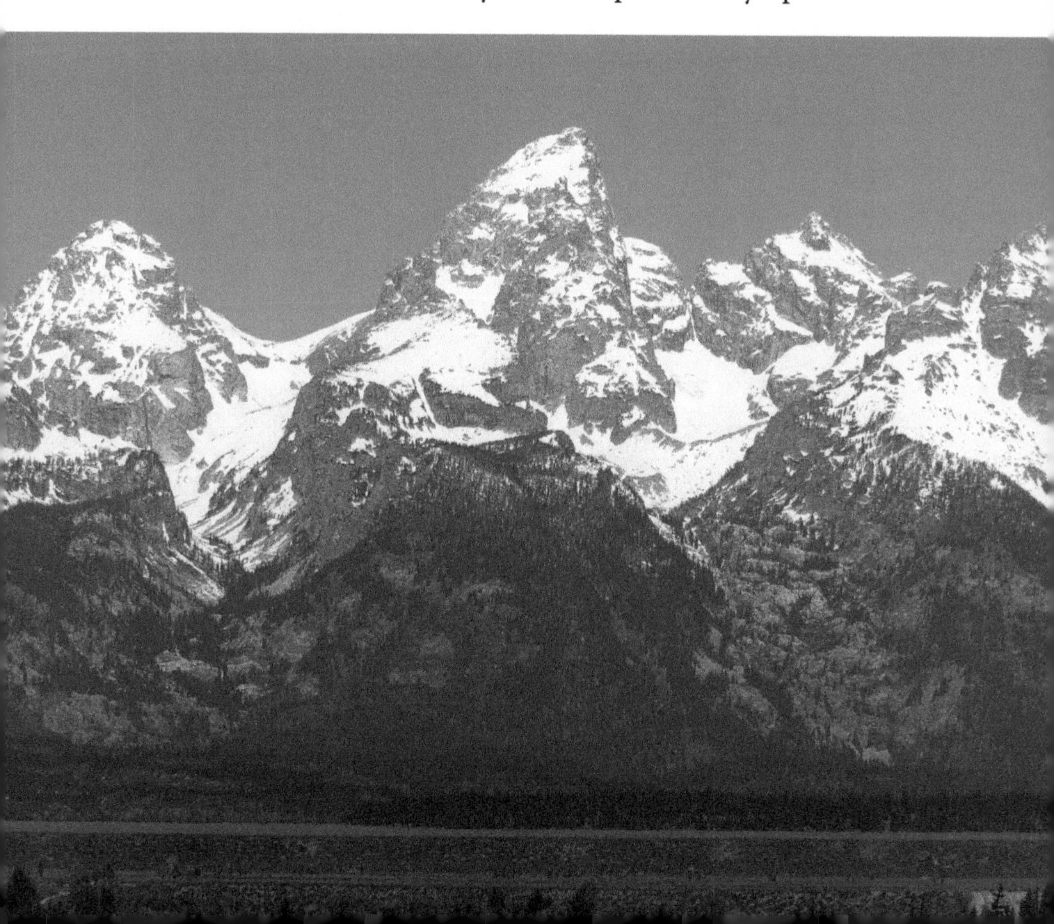

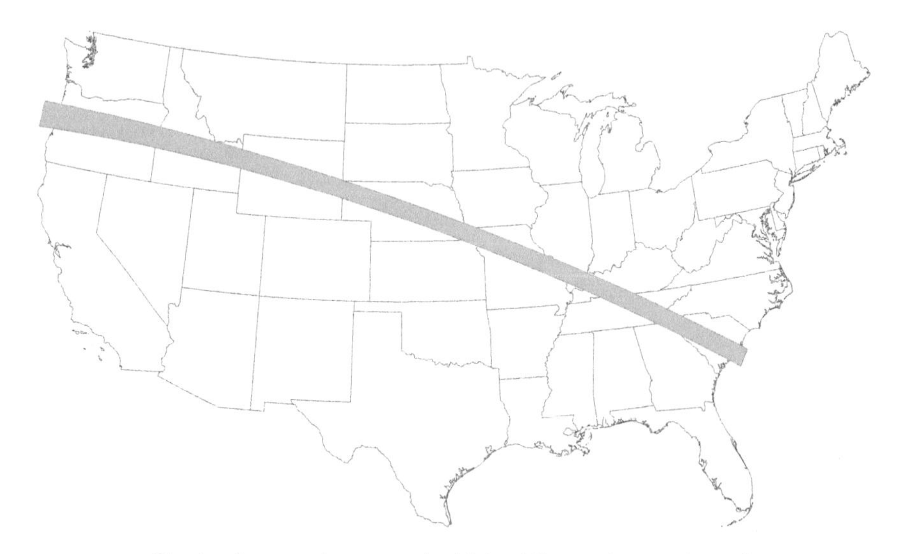

Total eclipse path across the United States (approximate).

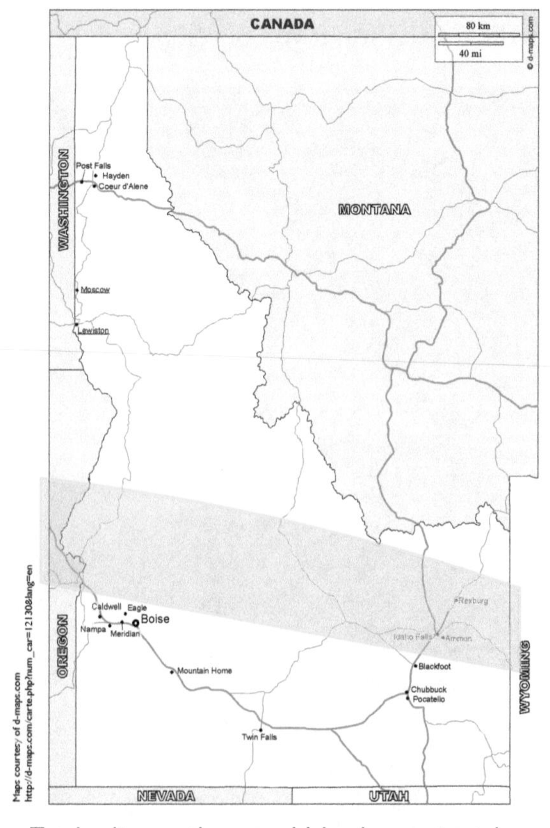

Total eclipse path across Idaho (approximate).

All About Eclipses

How an Eclipse Happens

An eclipse occurs when one celestial body falls in line with another, thus obscuring the sun from view. This occurs much more often than you'd think, considering how many bodies there are in the solar system. For instance, there are over 150 moons in the solar system. On Earth, we have two primary celestial bodies: the sun and the moon. The entire solar system is constantly in motion, with planets orbiting the sun and moons orbiting the planets. These celestial bodies often come into alignment. When these alignments cause the sun to be blocked, it is called an eclipse.

For an eclipse to occur, the sun, Earth, and moon must be in alignment. There are two types of eclipses: solar and lunar. A solar eclipse occurs when the moon obscures the sun. A lunar eclipse occurs when the moon passes through Earth's shadow. Solar eclipses are much more common, as we experience an average of 240 solar eclipses a century compared to an average of 150 lunar eclipses. Despite this, we are more likely to see a lunar eclipse than a solar eclipse. This is due to the visibility of each.

For a solar eclipse to be visible, you have to be in the moon's shadow. The problem with viewing a total eclipse is that the moon casts a small shadow over the world at any given time. You have to be in

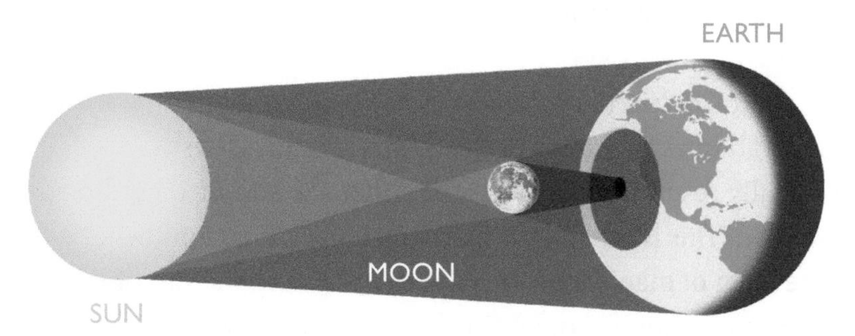

a precise location to view a total eclipse. The issue that arises is that most of these locations are inaccessible to most people. Though many would like to see a total solar eclipse, most aren't about to set sail for the middle of the Pacific Ocean. In fact, a solar eclipse is visible in the same place on the world on average every 375 years. This means that if you miss a solar eclipse above your hometown, you're not going to see another one unless you travel or move.

It's much easier to catch a glimpse of a lunar eclipse, even though they occur at a much lower frequency than their solar counterparts. A lunar eclipse darkens the moon for a few hours. This is different than a new moon when it faces away from the sun. During these eclipses, the moon fades and becomes nearly invisible.

Another result of a lunar eclipse is a blood moon. Earth's atmosphere bends a small amount of sunlight onto the moon turning it orange-red. The blood moon is caused by the dawn or dusk light being refracted onto the moon during an eclipse.

Lunar eclipses are much easier to see. Even when the moon is in the shadow of Earth, it's still visible throughout the world because of how much smaller it is than Earth.

Total vs. Partial Eclipse

What is the difference between a partial and total eclipse? A total eclipse of either the sun or the moon will occur only when the sun, Earth, and the moon are aligned in a perfectly straight line. This ensures that either the sun or the moon is partially or completely obscured.

In contrast, a partial eclipse occurs when the alignment of the three celestial bodies is not in a perfectly straight line. These types of eclipses usually result in only a part of either the sun or the moon being obscured. This is often what led to ancient civilizations believing that some form of magical beast or deity was eating the sun or the moon. It appears as though something has taken a bite out of either the sun or the moon during a partial eclipse.

Total eclipses, rarer than partial eclipses, still occur quite often. It's more difficult for people to be in a position to experience such an event firsthand. Total solar eclipses can only be viewed from a small portion of the world that falls into the darkest part of the moon's shadow. Often this happens in the middle of the ocean.

The Moon's Shadow

The moon's shadow is divided into two parts: the umbra and the penumbra. The former is much smaller than the latter, as the umbra is the innermost and darkest part of the shadow. The umbra is thus the central point of the moon's shadow, meaning that it is extremely small in comparison to the entire shadow. For a total solar eclipse to be visible, you need to be directly beneath the umbra of the moon's shadow. This is because that is the only point at which the moon completely blocks the view of the sun.

In contrast, the penumbra is the region of the moon's shadow in which only a portion of the light cast by the sun is obscured. When

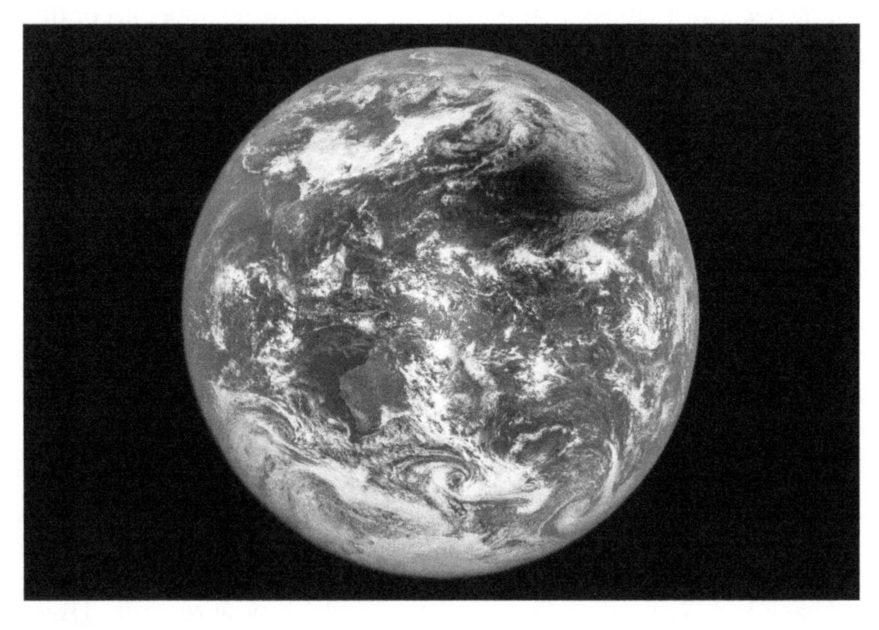

Total eclipse shadow 2016 as seen from 1 million miles on the Deep Space Climate Observatory satellite. Courtesy of NASA.

standing in the penumbra, you are viewing the eclipse at an angle. In the penumbra, the moon does not completely block the sun from view. This means that while the event is a total solar eclipse, you'll only see a partial eclipse. The umbra for the August 21st eclipse is approximately sixty miles wide. The penumbra will cover much of the United States.

To provide some context, the last total solar eclipse we experienced occurred on March 9, 2016, and was visible as a partial eclipse across most of the Pacific Ocean, parts of Asia, and Australia. However, the only place in the world to view this total solar eclipse was in a few parts of Indonesia.

Due to the varied locations and the brief periods for which they're visible, it's difficult to see each and every eclipse that occurs. Many people don't even realize that they have occurred. Consider that the umbra of the moon represents such a small fraction of the entire shadow and the majority of our planet is comprised of water. Thus, the rarity of being able to view a total solar eclipse increases significantly because it's likely that the umbra will fall over some part of the ocean rather than a populated landmass.

Eclipses Throughout History

Ancient peoples believed eclipses were from the wrath of angry gods, portents of doom and misfortune, or wars between celestial beings. Eclipses have played many roles in cultures, creating myths since the dawn of time. Both solar and lunar eclipses affected societies worldwide. Inspiring fear, curiosity, and the creation of legends, eclipses have cast a long shadow in the collective unconscious of humanity throughout history.

Early Myth & Astronomy

Documented observations of solar eclipses have been found as far back in history as ancient Egyptian and Chinese records. Timekeeping was important to ancient Chinese cultures. Astronomical observations were an integral factor in the Chinese calendar. The first observation of a solar eclipse is found in Chinese records from over 4,000 years ago. Evidence suggests that ancient Egyptian observations may predate those archaic writings.

Many ancient societies, including Roman, Greek and Chinese civilizations, were able to infer and foresee solar eclipses from astronomical data. The sudden and unpredictable nature of solar eclipses had a stressful and intimidating effect on many societies that lacked the scientific insight to accurately predict astronomical events. Relying on the sun for their agricultural livelihood, those societies interpreted solar eclipses as world-threatening disasters.

In ancient Vietnam, solar eclipses were explained as a giant frog eating the sun. The peasantry of ancient Greece believed that an eclipse was the sign of a furious godhead, presenting an omen of wrathful retribution in the form of natural disasters. Other cultures were less speculative in their investigations. The Chinese Song Dynasty scientist Shen Kuo proved the spherical nature of the Earth and heavenly bodies through scientific insight gained by the study of eclipses.

The Eclipse in Native American Mythology

Eclipses have played a significant role in the history of the United States. Before Europeans settled in the Americas, solar eclipses were important astronomical events to Native American cultures. In most native cultures, an eclipse was a particularly bad omen. Both the sun and the moon were regarded as sacred. Viewing an eclipse, or even being outside for the duration of the event, was considered highly taboo by the Navajo culture. During an eclipse, men and women would simply avert their eyes from the sky, acting as though it was not happening.

The Choctaw people had a unique story to explain solar eclipses. Considering the event as the mischievous actions of a black squirrel and its attempt to eat the sun, the Choctaw people would do their best to scare away the cosmic squirrel by making as much noise as possible until the end of the event, at which point cognitive bias would cause them to believe they'd once again averted disaster on an interplanetary scale.

Contemporary American Solar Phenomena

The investigation of solar phenomena in twentieth-century American history had a similarly profound effect on the people of the United States. A total solar eclipse occurring on the sixteenth of June, 1806, engulfed the entire country. It started near modern-day Arizona. It passed across the Midwest, over Ohio, Pennsylvania, New York, Massachusetts, and Connecticut. The 1806 total eclipse was notable for being one of the first publicly advertised solar events. The public was informed beforehand of the astronomical curiosity through a pamphlet written by Andrew Newell entitled *Darkness at Noon, or the Great Solar Eclipse.*

This pamphlet described local circumstances and went into great detail explaining the true nature of the phenomenon, dispelling myth and superstition, and even giving questionable advice on the best methods of viewing the sun during the event. Replete with a short historical record of eclipses through the ages, the *Darkness at Noon* pamphlet is one of the first examples of an attempt to capitalize on the mysterious nature of solar eclipses.

Another notable American solar eclipse occurred on June 8, 1918. Passing over the United States from Washington to Florida, the eclipse was accurately predicted by the U.S. Naval Observatory and heavily documented in the newspapers of the day. Howard Russell Butler, painter and founder of the American Fine Arts Society, painted the eclipse from the U.S. Naval Observatory, immortalizing the event in *The Oregon Eclipse.*

Four more total solar eclipses occurred over the United States in the years 1923, 1925, 1932, and 1954, with another occurring in 1959. The October 2, 1959, solar eclipse began over Boston, Massachusetts. It was a sunrise event that was unviewable from the ground level. Eminent astronomer Jay Pasachoff attributed this event to sparking his interest in the study of astronomy. Studying under Professor Donald Menzel of Williams College, Pasachoff was able to view the event from an airline hired by his professor.

To this day, many myths surround the eclipse. In India, some local customs require fasting. In eastern Africa, eclipses are seen as a danger to pregnant women and young children. Despite the mystery and legend associated with unique and rare astronomical events, eclipses continue to be awe-inspiring. Even in the modern day, eclipses draw out reverential respect for the inexorable passing of celestial bodies. They are a reminder of the intimate relationship between the denizens of Earth and the universe at large.

Present Day Eclipses

The year 2016 brought the world just two solar eclipses. A total solar eclipse occurred on the 9th of March. An annular solar eclipse, in which the sun appears as a "ring of fire" occurred on the 23rd of March. If you're interested in seeing this rare and exciting solar

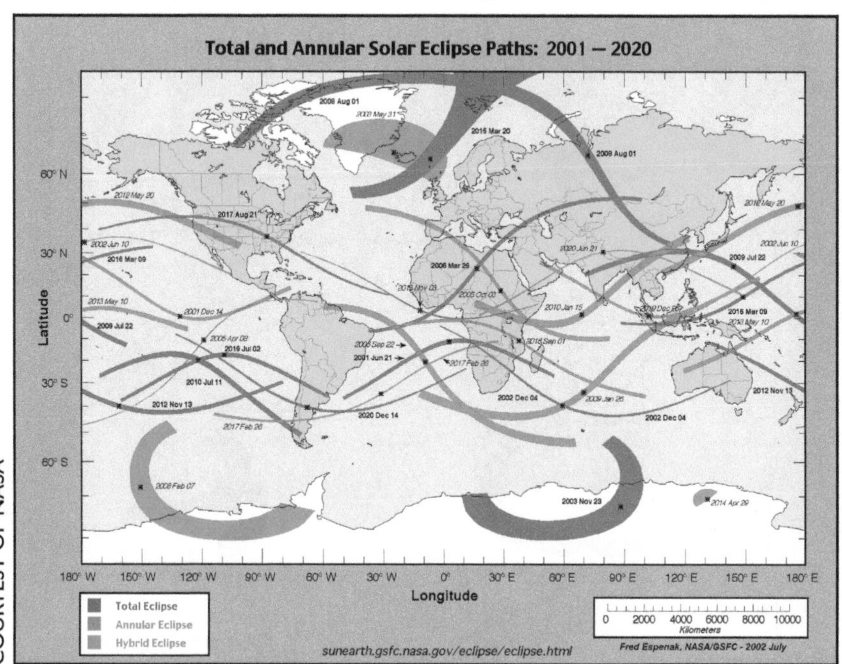

COURTESY OF NASA

phenomenon yourself, you must to travel to either South America or Western Africa on the 26th of February, 2017.

The next total solar eclipse viewable from the United States, or anywhere else in the world, will occur on the 21st of August, 2017. It will be visible in Oregon, Idaho, Wyoming, Nebraska, Missouri, Kentucky, Tennessee, and South Carolina. The event will be the only total solar eclipse for Americans this decade.

Future American Eclipses

The next total eclipse to cross the continental United States is April 8, 2024. It will travel from Texas to Maine. After that, the next American total eclipses will be in 2044 and 2045. The next partial eclipse to touch Idaho will be in 2046. You will have to wait until 2169 for the next total eclipse to pass over southwestern Idaho. That is as far into the future as the end of the Civil War was in the past. In 2252, a total eclipse on December 31 will close out the year in Idaho.

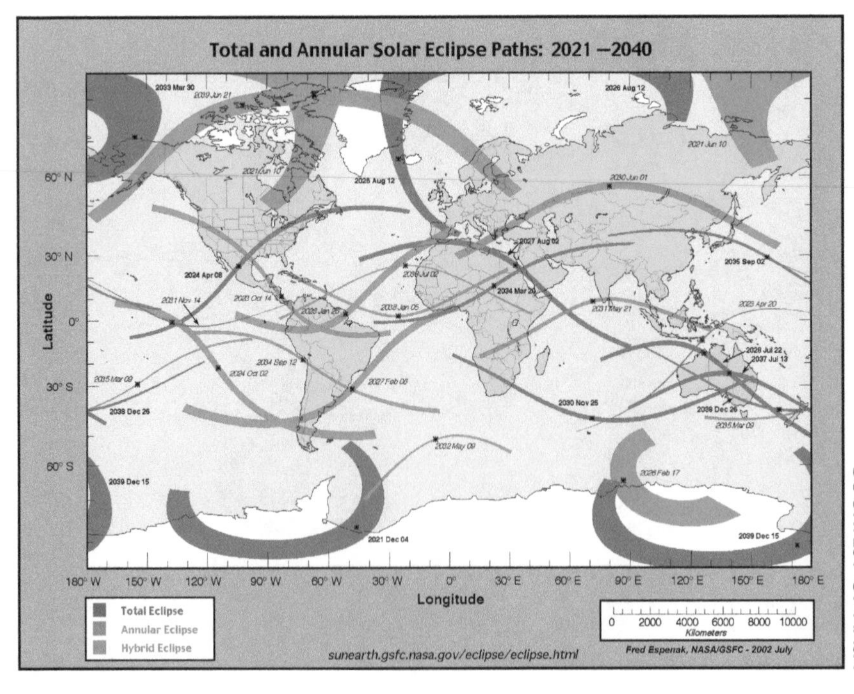

COURTESY OF NASA

Viewing and Photographing the Eclipse

At-home Pinhole Method

Use the pinhole method to view the eclipse safely. It costs little but is the safest technique there is. Take a stiff piece of single-layer cardboard and punch a clean pinhole. Let the sun shine through the pinhole onto another piece of cardboard. That's it!

Never look at the sun through the pinhole. Your back should be toward the sun to protect your eyes. To brighten the image, simply move the back piece of cardboard closer to the pinhole. To see it larger, move the back cardboard farther away. Do not make the pinhole larger. It will only distort the crescent sun.

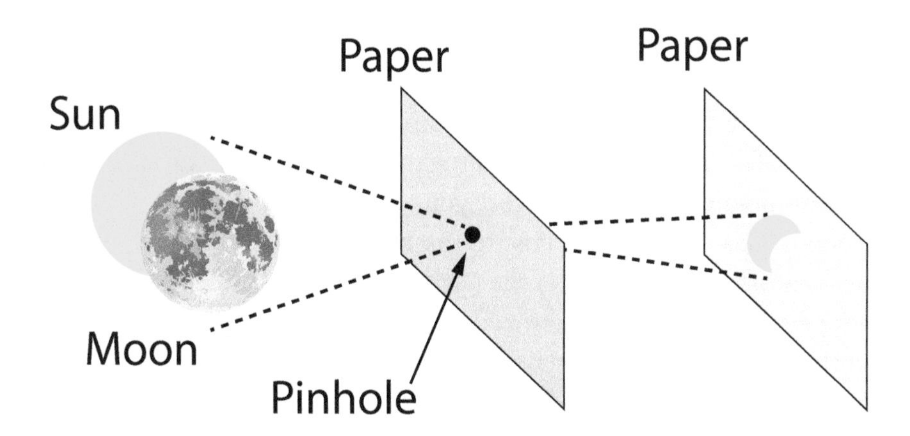

Welding Goggles

Welding goggles that have a rating of fourteen or higher are another useful eclipse viewing tool. The goggles can be used to view the solar eclipse directly. Do not use the goggles to look through binoculars or telescopes, as the goggles could potentially shatter due to intense direct heat. Avoid long periods of gazing with the goggles. Look away every so often. Give your eyes a break.

Solar Filters for Telescopes

The ONLY safe way to view solar eclipses using telescopes or binoculars is to use solar filters. The filters are coated with metal to diminish the full intensity of the sun. Although the filters can be expensive, it is better to purchase a quality filter rather than an inexpensive one that could shatter or melt from the heat.

The filters attach to the front of the telescope for easy viewing. Remember to give your telescope cooling breaks. Rapid heating can damage your equipment with or without filters attached.

Watch Out for Unsafe Filters

There are several myths surrounding solar filters for eclipse viewing. In order for filters to be safe, they must be specially designed for looking at a solar eclipse. The following are all unsafe for eclipse viewing and can lead to retinal damage: developed colored or chromogenic film, black-and-white negatives such as X-rays, CDs with aluminum, smoked glass, floppy disk covers, black-and-white film with no silver, sunglasses, or polarizing films.

Some online articles state that using developed black-and-white film is safe. Those articles fail to mention the film must have a layer of real metallic silver to protect your eyes. Using developed film is discouraged. You cannot ensure the quality of the film. Feeling no discomfort while looking at the partial solar eclipse does NOT mean your eyes are protected. Retinal damage can occur with zero pain due to the retinas having no pain receptors. Please be careful. Only use protective glasses certified for viewing the eclipse.

Viewing with Binoculars

When viewing the eclipse with binoculars, it is important to use solar filters on both lenses until totality. Only then is it safe to remove the filter. As the sun becomes visible after totality, replace the filters for safe viewing. Protect your pupils. Remember to give your binoculars a cool-down break between viewings. They can overheat rapidly from being pointed directly at the sun even with filters attached.

Planning Ahead

There are many things to keep in mind when viewing a total eclipse. It is important to plan ahead to get the most out of this extraordinary experience.

Understanding Sun Position

All compass bearings in this book are true north. All compasses point to Earth's magnetic north. The difference between these two measurements is called magnetic declination. The magnetic declination for eastern Idaho in August 2017 is:

11° 59' (for Idaho Falls)

Subtract the declination from the azimuth bearing as given in the text, and set your compass to that direction.

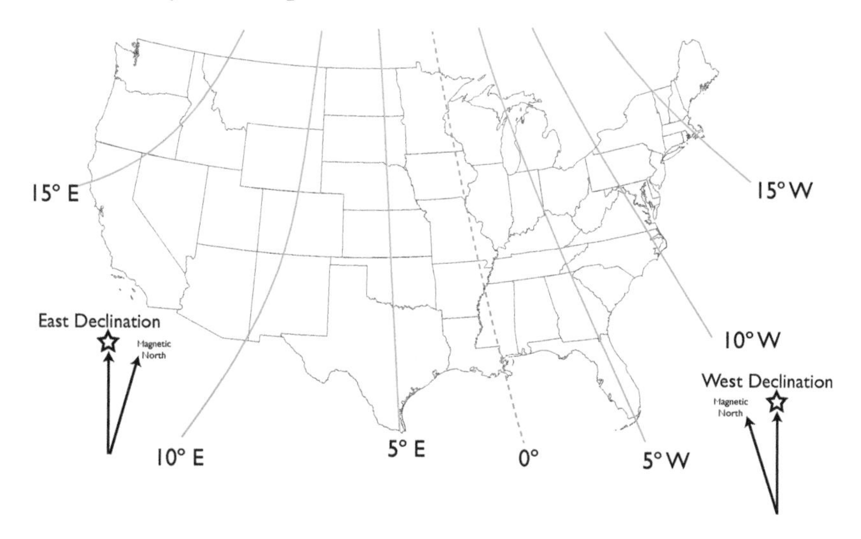

If you purchase a compass with a built-in declination adjustment, you can change the setting once and eliminate the calculations. The Suunto M-3G compass has this correction. A compass with a sighting mirror or wire will help you make a more accurate azimuth sighting.

The Suunto M-3G also has an inclinometer. This allows you to measure the elevation of any object above the horizon. Use this to figure out how high the sun will be above your position.

You can also use a smartphone inclinometer and compass for this purpose. Make sure to calibrate your smartphone's compass before every use, otherwise it might indicate the wrong bearing. Set the smartphone compass for true north to match the book. Understand the compass prior to August 21. There will be little time to guess or search on Google. Smartphone and GPS compasses are "sticky." Their compasses don't swing as freely as a magnetic compass does.

The author has used his magnetic compass for azimuth measurements and a smartphone to measure elevation. Combining these two tools will allow you to make the best sightings possible.

Outdoor sporting goods stores in most Idaho towns and cities carry compasses. We recommend purchasing a good compass in your hometown. Take the time to learn how to use it before the day of the eclipse. You do not want to struggle with orienteering basics under pressure.

Sun Azimuth

Azimuth is the compass angle along the horizon, with 0° corresponding to north, and increasing in a clockwise direction. 90° is east, 180° is south, and 270° is west.

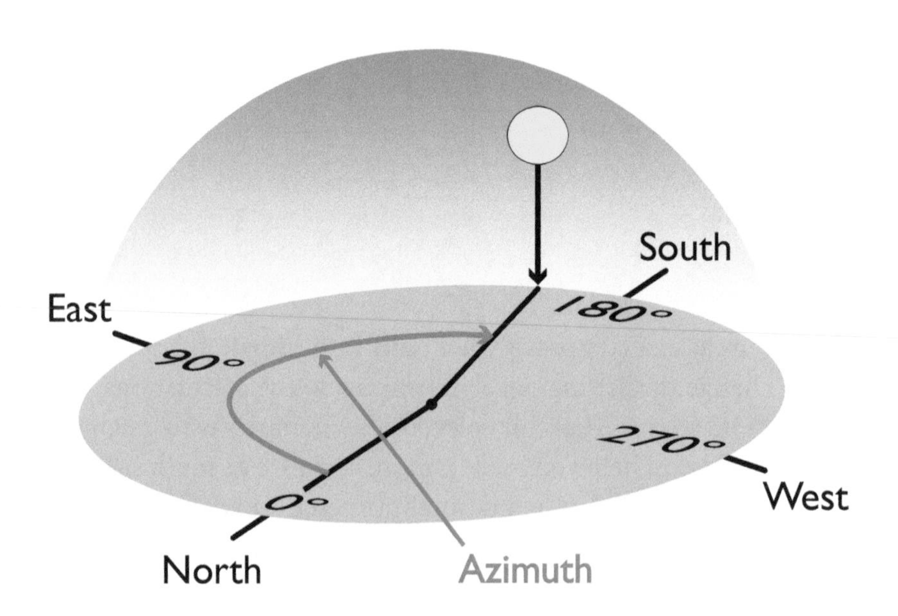

Sun Elevation

Altitude is the sun's angle up from the horizon. A 0° altitude means exactly on the horizon and 90° means "straight up."

Using the sun azimuth and elevation data, you can predict the position of the sun at any given time. Positions given in this book coincide with the time of eclipse totality unless otherwise noted.

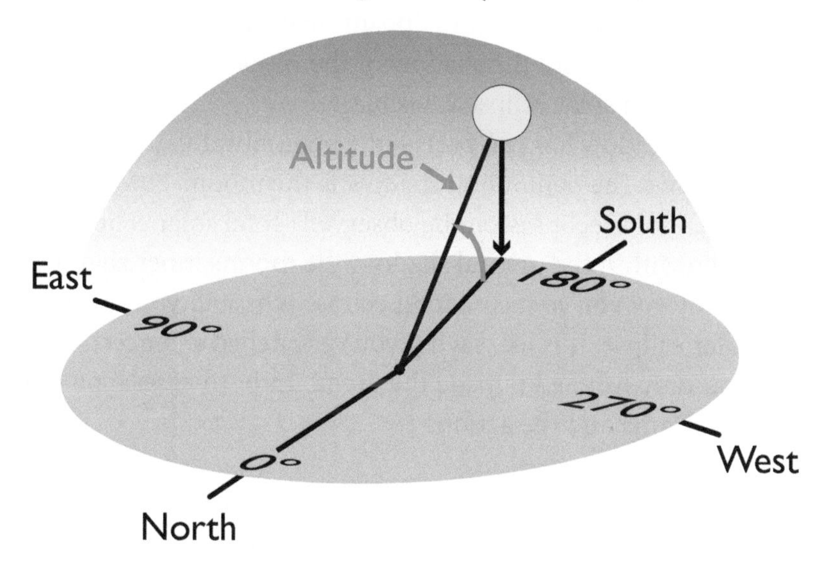

Eclipse Data for Select Idaho Locations

EVENT	TOTALITY START (MDT)	ALTITUDE	AZIMUTH
DRIGGS	11:34:20AM	50°	134°
IDAHO FALLS	11:32:59AM	49°	132°
MACKAY	11:30:19AM	48°	130°
REXBURG	11:33:13AM	49°	133°
SMITHS FERRY	11:26:37AM	46°	126°
SUN VALLEY	11:29:36AM	47°	128°

Eclipse Photography

Photographing an eclipse is an exciting challenge, as the moon's shadow moves near 2,000MPH. There is an element of danger and the pressure of time. Looking at the unfiltered sun through a camera can permanently damage your vision and your camera. If you are unsure, just enjoy the eclipse with specially designed glasses. Keep a solar filter on your lens during the eclipse and remove for the duration of totality!

Partial Vs. Total Solar Eclipse

To successfully and safely photograph a partial and total eclipse, it is important to understand the difference between the two. A solar eclipse occurs when the moon is positioned between the sun and Earth. The region where the shadow of the moon falls upon Earth's surface is where a solar eclipse is visible.

The moon's shadow has two parts—the penumbral shadow and the umbral shadow. The penumbral shadow is the moon's outer shadow where partial solar eclipses can be observed. Total solar eclipses can only be seen within the umbral shadow, the moon's inner shadow.

You cannot say you've seen a total eclipse when all you saw was a partial solar eclipse. It is like saying you've watched a concert, but in reality, you only listened outside the arena. In both cases, you have missed the drama and the action.

Photographing A Partial And Total Solar Eclipse

Aside from the region where the outer shadow of the moon is cast, a partial solar eclipse is also visible before a total solar eclipse within

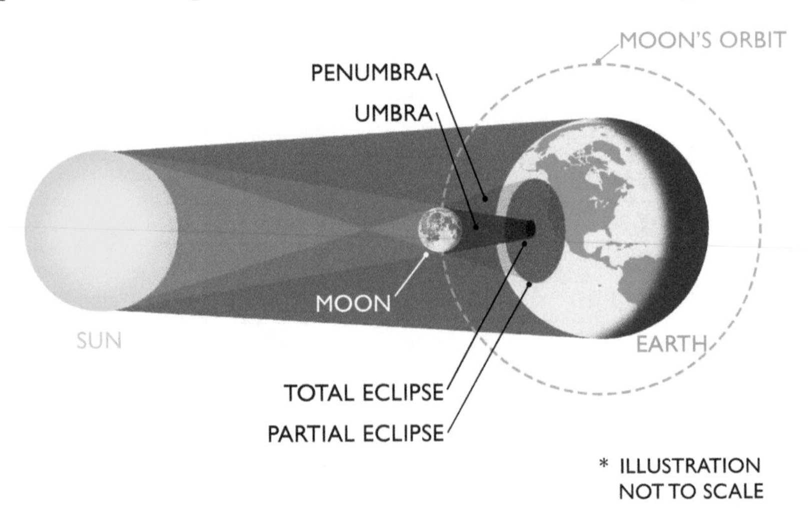

the inner shadow region. In both cases, it is imperative to use a solar filter on the lens for both photography and safety reasons. This is the only difference between taking a partial eclipse and a total eclipse photograph of the sun.

To photograph a total solar eclipse, you must be within the Path of Totality, the surface of the Earth within the moon's umbral shadow.

The Challenge

A total solar eclipse only lasts for a couple of minutes. It is brief, but the scenario it brings is unforgettable. Seeing the radiant sun slowly being covered by darkness gives the spectator a high level of anticipation and indescribable excitement. Once the moon completely covers the sun's radiance, the corona is finally visible. In the darkness, the sun's corona shines, capturing the crowd's full attention. Watching this phenomenon is a breathtaking experience.

Amidst all the noise, cheering, and excitement, you have less than 150 seconds to take a perfect photograph. The key to this is planning. You need to plan, practice, and perfect what you will do when the big moment arrives because there is no replay. The pressure is enormous. You only have two minutes to capture the totality and the sun's corona using different exposures.

Plan, Practice, Perfect

It is important to practice photographing before the actual phenomenon arrives. Test your chosen imaging setup for flaws. Rehearse over and over until your body remembers what you will do from the moment you arrive at your chosen spot to the moment you pack up and leave the area.

You will discover potential problems regarding vibrations and focus that you can address immediately. This minimizes the variables that might affect your photographs at the most critical moment.

It's common for experienced eclipse chasers to lose track of what they plan to do. Write down what you expect to do. Practice it time and again. Play annoying, distracting music while you practice. Try photographing in the worst weather possible. Do anything you can to practice under pressure. Eclipse day is not the time to practice.

Once the sun is completely covered, don't just take photographs. Capture the experience and the image of the total solar eclipse in your mind as well. Set up cameras around you to record not just the total solar eclipse but also the excitement and reaction of the crowd.

Eclipse Photography Gear

What do you need to photograph the total eclipse? There are only a few pieces of equipment that you'll need. Preparing to photograph an eclipse successfully takes time. Not only do you have to be skilled and have the right gear, you have to be in the correct place.

Basic Eclipse Photography Equipment

- Solar viewing glasses
- Lens solar filter
- Minimum 300mm lens
- Stable tripod that can be tilted to 60° vertical
- High-resolution DSLR
- Spare batteries for everything
- Secondary camera to photograph people, the horizon, etc.
- Remote cable or wireless release

Additional Items

- Video camera
- Video camera tripod
- Quality pair of binoculars
- Solar filters for each binocular lens
- Photo editing software

Equipment to Prepare Before the Big Day

A. Solar viewing glasses

You need a pair of solar viewing glasses as the eclipse approaches.

B. Solar Filter

Partial and total eclipse photography is different from normal photography. Even if only 1% of the sun's surface is visible, it is still approximately 10,000 times brighter than the moon. Before totality, use a solar filter on your lens. Do not look at the sun with your eyes. It can cause irreparable damage to your retinas.

DO NOT leave your camera pointed at the sun without a solar filter attached. The sun will melt the inside of your camera. Think of a magnifying glass used to torch ants and multiply that by one hundred.

C. Lens

To capture the corona's majesty, you need to use a telescope or a telephoto lens. The best focal length, which will give you a large image of the sun's disk, is 400mm and above. You don't want to waste all your efforts by bringing home a small dot where the black disk and majestic corona are supposed to be.

D. Tripod

Bring a stable enough tripod to support your camera properly to avoid unsteady shots and repeated adjustments. Either will ruin your photos. It also needs to be portable in case you need to change locations for a better shot.

E. Camera

You need to remember to set your camera to its highest resolution to capture all the details. Set your camera to:

- 14-bit RAW is ideal, otherwise
- JPG, Fine compression, Maximum resolution

Bracket your exposures. Shoot at various shutter speeds to capture different brightnesses in the corona. Note that stopping your lens all the way down may not result in the sharpest images.

Choose the lowest possible ISO for the best quality while maintaining a high shutter speed to prevent blurred shots. Set your camera to manual. Do not use AUTO ISO. Your camera will be fooled. The night before, test the focus position of your lens using a bright star or the moon.

Constantly double-check your focus. Be paranoid about this. You can deal with a grainy picture. No amount of Photoshop will fix a blurry, out-of-focus picture.

F. Batteries

Remember to bring fresh batteries! Make sure that you have enough power to capture the most important moments. Swap in fresh batteries thirty minutes before totality.

G. Remote release

Use a wired or wireless remote release to fire the camera's shutter. This will reduce the amount of camera vibration.

H. Video Camera

Run a video camera of yourself. Capture all the things you say and do during the totality. You'll be amazed at your reaction.

I. Photo editing software

You will need quality photo editing software to process your eclipse images. Adobe Lightroom and Photoshop are excellent programs to extract the most out of your images. Become well versed in how to use them at least a month before the eclipse.

J. Smartphone applications

The following smartphone applications will aid in your photography planning: Wunderground, Skyview, Photographer's Ephemeris, Sunrise and Sunset Calculator, SunCalc, and Sun Surveyor among others.

Camera Phones

Smartphone cameras are useful for many things but not eclipse photography. An iPhone 6 camera has a 63° horizontal field of view and is 3264 pixels across. If you attempt to photograph the eclipse, the sun will be a measly 30-40 pixels wide depending on the phone. Digital pinch zoom won't help here. If you want *National Geographic* images, you'll need a serious camera and lens, far beyond any smartphone.

Consider instead using a smartphone to run a time-lapse of the entire event. The sun will be minuscule when shot on a smartphone. Think of something else exciting and interesting do to with it. Purchase a Gorilla Pod, inexpensive tripod, or selfie stick and mount the smartphone somewhere unique.

Also, partial and total eclipse light is strange and ethereal. Consider using that light to take unique pictures of things and people. It's rare and you may have something no one else does.

Focal Length & the Size of Sun

The size of the sun in a photo depends on the lens focal length. A 300mm lens is the recommended minimum on a full-frame (FF) DSLR. Lenses up to this size are relatively inexpensive. For more magnification, use an APS-C (crop) size sensor. Cameras with these sensors provide an advantage by capturing a larger sun.

For the same focal length, an APS-C sensor will provide a greater apparent magnification of any object. As a consequence, a shorter, less expensive lens can be used to capture the same size sun.

The below figure shows the size of the sun on a camera sensor at various focal lengths. As can be seen with the 200mm lens, the sun is quite small. On a full-frame camera at 200mm, the sun will be 371 pixels wide on a Nikon D810, a 36-megapixel body. A lower resolution FF camera will result in an even smaller sun.

Printing a 24-inch image shot on a Nikon D810 with a 200mm lens at a standard 300 pixels per inch results in a small sun. On this size paper, the sun will be a miserly 1.25 inches wide!

Photographing the eclipse with a lens shorter than 300mm will leave you with little to work with. Using a 400mm lens and printing a 24-inch print will result in a 2.5-inch-wide sun. For as massive as the sun is, it is a challenge to take a photograph with the sun of any meaningful size.

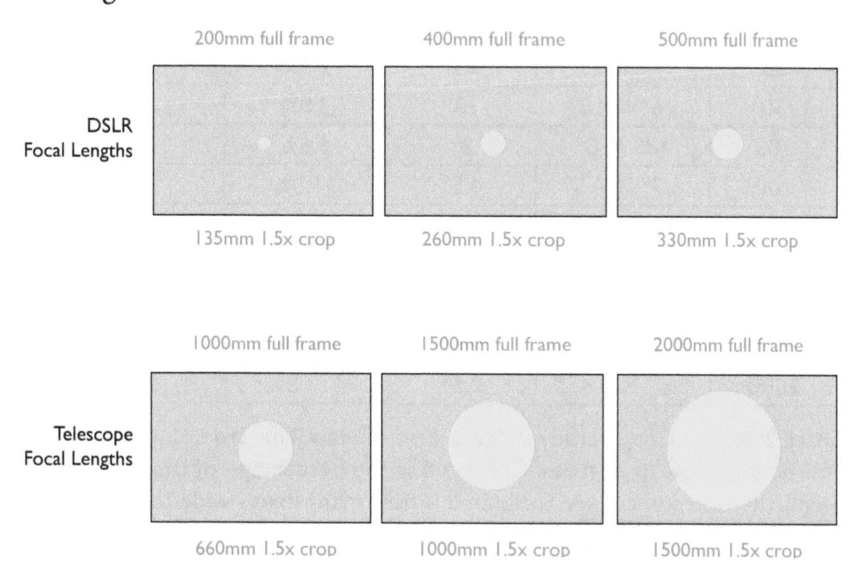

FOCAL LENGTH	FOV FULL FRAME	FF VERT. ANGLE	% OF FF	SUN PIXEL SIZE
14	104° X 81°	81°	0.7%	32.1
20	84° X 62°	62°	0.9%	41.9
28	65° X 46°	46°	1.2%	56.5
35	54° X 38°	38°	1.4%	68.5
50	40° X 27°	27°	2.0%	96.4
105	19° X 13°	13°	4.1%	200.2
200	10° X 7°	7°	7.6%	371.9
400	5° X 3.4°	3.4°	15.6%	765.6
500	4° X 2.7°	2.7°	19.6%	964.2
1000	2° X 1.3°	1.3°	40.8%	2002.5
1500	1.4° X 0.9°	0.9°	58.9%	2892.6
2000	1° X 0.68°	0.68°	77.9%	3828.4

Chart 1: Full-frame camera field of view. The 3rd column is the vertical field of view in degrees. Column 4 is the percentage of the total sensor height that the sun covers. Column 5 is how many pixels wide the sun will be on a 36MP Nikon D810. (Values are estimates)

FOCAL LENGTH	FOV CROP	CROP VERT DEG	% OF CROP	SUN PIXEL SIZE
14	80° X 58°	58°	0.9%	33.9
20	61° X 43°	43°	1.2%	45.8
28	45° X 31°	31°	1.7%	63.5
35	37° X 25°	25°	2.1%	78.7
50	26° X 18°	18°	2.9%	109.3
105	13° X 8°	8°	6.6%	245.9
200	6.7° X 4.5°	4.5°	11.8%	437.2
400	3.4° X 2°	2°	26.5%	983.7
500	2.7° X 1.8	1.8°	29.4%	1093.0
1000	1.3° X 0.9°	0.9°	58.9%	2186.0
1500	0.9° X 0.6°	0.6°	88.3%	3278.9
2000	0.6° X 0.45°	0.5°	117.8%	4371.9

Chart 2: APS-C Crop sensor camera field of view. The 3rd column is the vertical field of view in degrees. Column 4 is the percentage of the total sensor height that the sun covers. Column 5 is how many pixels wide the sun will be on a 12mp Nikon D300s. (Values are estimates)

The big challenge is the cost of the lens. Lenses longer than 300mm are expensive. They also require heavier tripods and specialized tripod heads. The 70-300mm lenses from Nikon, Canon, Tamron, and others are relatively affordable options. It is worth spending time at a local camera shop to try different lenses. Long focal-length lenses are a significant investment, especially for a single event.

To achieve a large eclipse image, you will need a long focal-length lens, ideally at least 400mm. A standard 70-300mm lens set to 300mm will show a small sun. At 500mm, the sun image becomes larger and covers more of the sensor area. The corona will take up a significant portion of the frame. By 1000mm, the corona will exceed the capture area on a full-frame sensor. See the picture on page thirty-seven for sun size simulations for different focal lengths.

Suggested Exposures

To photograph the partial eclipse, the camera must have a solar filter attached. If not, the intense light from the sun may damage (fry) the inside of your camera. This has happened to the author. The exposure depends on the density (darkness) of the solar filter used.

As a starting point, set the camera to ISO 100, f/8, and with the solar filter on, try an exposure of 1/4000. Make adjustments based on the histogram and highlight warning.

Turn on the highlight warning in your camera. This feature is commonly called "blinkies." This warning will help you detect if the image is overexposed or not.

Once the Baily's Beads, prominences, and corona become visible, there will only be 2.25 minutes to take bracketed shots. It will take at least eleven shots to capture the various areas of the sun's corona. The brightness varies considerably. No commercially available camera can capture the incredible dynamic range of the different portions of the delicate corona. This requires taking multiple photographs and digitally combining them afterward.

During totality, try these exposure times at ISO 100 and f/8:

1/4000, 1/2000, 1/1000, 1/250, 1/60, 1/30, 1/15, 1/4, 1/2, 1 sec, and 4 sec.

Photography Time

Set the camera to full-stop adjustments. It will reduce the time spent fiddling. As an example, the author tried the above shot sequence, adjusting the shutter speed as fast as possible.

It took thirty-three seconds to shoot the above 11 shots using 1/3-stop increments. This was without adjusting composition, focus, or anything else but the shutter speed. When the camera was set to full stop increments, it only took twenty-two seconds to step through the same shutter speed sequence.

Assuming the totality lasts less than two minutes, only four shot sequences could be made using 1/3-stop increments. Yet six shot sequences could be made when the camera was set to full stop steps. Zero time was spent looking at the back LCD to analyze highlights and the histogram.

Now add in the bare minimum time to check the highlight warning. It took sixty-three seconds to shoot and check each image using full stops. And that was without changing the composition to allow for sun movement, bumping the tripod, etc. Looking at the LCD ("chimping") consumed **half** of the totality time.

This test was done in the comfort of home under no pressure. In real world conditions, it may be possible to successfully shoot only one sequence. If you plan to capture the entire dynamic range of the totality, you must practice the sequence until you have it down cold. If you normally fumble with your camera, do not underestimate the difficulty, frustration, and stress of total eclipse photography.

Most importantly, trying to shoot this sequence allowed for zero time to simply look at the totality to enjoy the spectacle.

Avoid Last Minute Purchases

You should purchase whatever you think you'll need to photograph the eclipse today. This event will be nothing short of massive. Remember the hot toy of the year? Multiply that frenzy by a thousand. Everyone will want to try to capture their own photo.

Do not wait until the last few weeks before the eclipse to purchase cameras, lenses, filters, tripods, viewing glasses, and associated material. Consider that the totality of the eclipse will streak from coast to coast. Everyone who wants to photograph the eclipse will order at the same time. If you wait until August to buy what you need, it's conceivable that every piece of camera equipment capable of creating a total eclipse photo will be sold out in the United States. Whether this happens or not, do not wait until midsummer to make your purchases. It may be too late.

Practice

You will need to practice with your equipment. Things may go wrong that you don't anticipate. If you've never photographed a partial or total eclipse, taking quality shots is more difficult than you think. Practice shooting the sequence with a midday sun. This will tell you if you have your exposures and timing correct. Figure out what you need well in advance.

Practice photographing the full moon and stars at night. Capture the moon in full daylight. There will be six moon cycles to practice with. Astrophotography is challenging and requires practice.

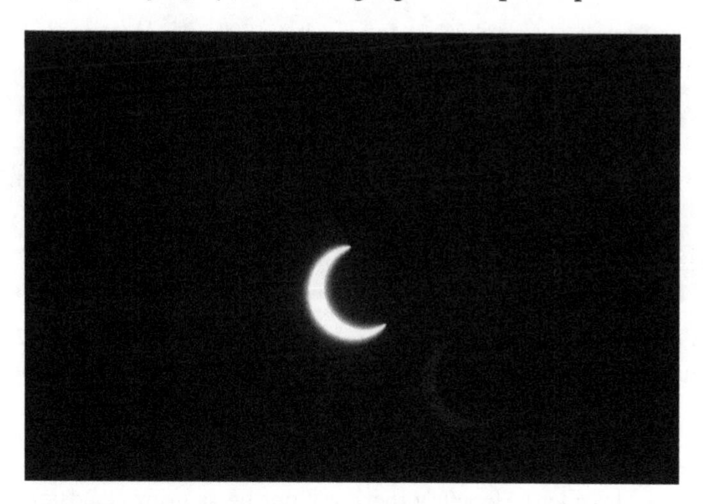

The May 20, 2012, eclipse as seen in San Diego, CA, shot with a Nikon D300s (crop sensor) with an 80-400mm lens set to ~350mm. The sun is 560 pixels wide on the 4288x2848 image.

This image is shown straight out of the camera without modification. Even with a high-quality camera and lens, photographing an eclipse is challenging. Note the haze and reflection from the overexposed sun.

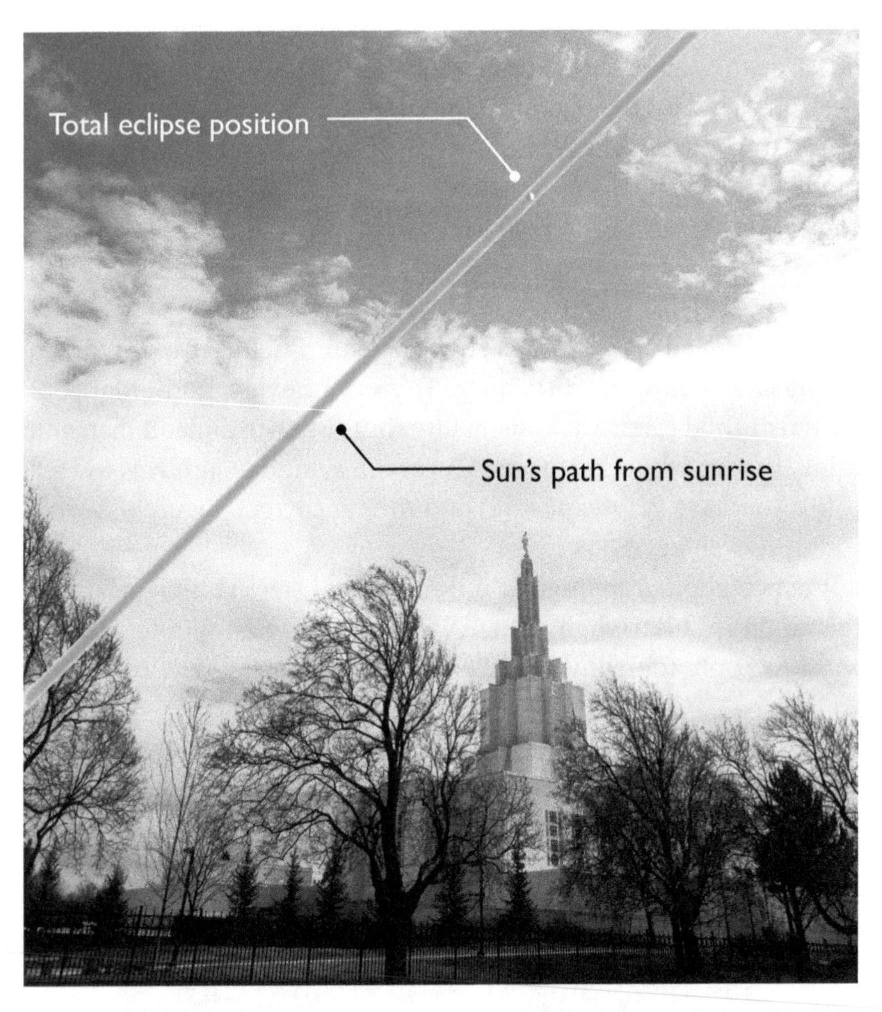

The sun will follow this path on the morning of the eclipse on August 21, 2017. The shot was taken from outside the fence on Riverside Drive near the Snake River.

This image of the Idaho Falls Idaho Mormon Temple was created with a Nikon D800 and a 24mm lens. Hopefully there will not be as many clouds on the day of the eclipse.

⊙ is the symbol for the sun and first appeared in Europe during the Renaissance.

☾ is the ancient symbol for the moon.

Viewing Locations Around Idaho

Hundreds of thousands of people will visit Idaho to view the total eclipse. Most will visit a few cities and towns in eastern Idaho along the I-15 corridor. Due to the position of the sun when the eclipse happens, capturing the eclipse near classic mountain summits will be possible in east Idaho. The sun will be in the southeast sky at the moment of totality across the state.

Most of the famous viewpoints in Driggs, Victor, Tetonia, and Lamont areas will allow viewers to see the total eclipse over the Tetons. Visitors to Jackson will be surprised to find out they will not be able to have the same eclipse view over the Tetons as visitors to Idaho.

This section contains popular, alternative, and little-known locations to watch the eclipse. As long as there are no clouds or smoke from fires, the partial eclipse will be viewable from anywhere in Idaho.

Suggested Total Eclipse View Points

Towns and Cities

- Arco
- Cascade
- Driggs
- Howe & Berenice
- Idaho Falls
- Lamont
- Mackay
- Midvale
- Mud Lake & Terreton
- Rexburg
- Rigby
- Smiths Ferry
- Stanley
- Sun Valley & Ketchum
- Tetonia
- Weiser

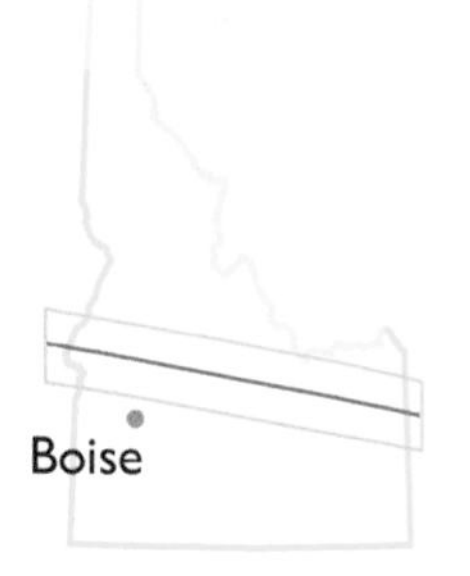

Unique Locations

- Borah Peak
- Mann Creek State Park
- Menan Buttes
- Salmon-Challis NF
- St. Anthony Sand Dunes

Arco

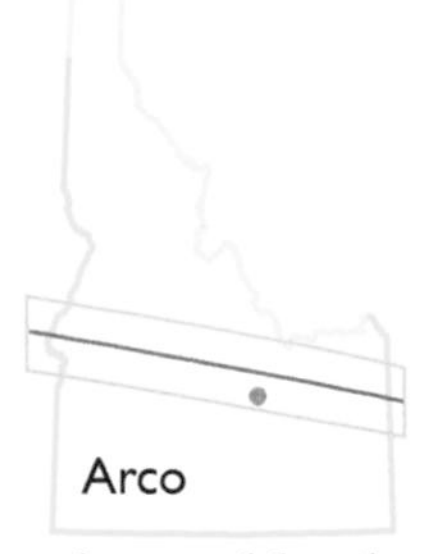

Elevation:	5,325 feet
Population:	910 (2013)
Main road/hwy:	US 26/93

Overview

Arco holds the distinction of being the first town in the world to be lit with electricity generated by nuclear power. Nearby Atomic City is the site of the world's first nuclear reactor. Both make for a unique total eclipse-viewing experience. The town will have a solar eclipse party at Butte County High School, a morning event at Bottolfsen Park, a solar telescope at the town visitor center, and Arco's Rockin' Country Outdoors music festival. Drive through to visit Craters of the Moon National Monument, a surreal volcanic landscape, to watch the total eclipse from. Note: only the northernmost corner of the national monument will be in the totality.

Getting There

Drive west from Idaho Falls on US 20/26 for sixty-seven miles to reach Arco.

Totality Duration

1 minutes 41 seconds

Notes

Visit this website for more information about Arco's eclipse events: arcosrockincountryoutdoors.com/come-watch-the-eclipse/

Event	Time (MDT)	Altitude	Azimuth
Sunrise	6:45:00AM	0°	73°
Eclipse Start	10:13:47AM	36°	110°
Totality Start	11:31:03AM	48°	130°
Totality End	11:32:40AM	49°	131°
Eclipse End	12:55:35PM	57°	162°
Sunset	8:26:00PM	0°	287°

Cascade

Elevation: 4,760 feet
Population: 904 (2013)
Main road/hwy: Hwy 55

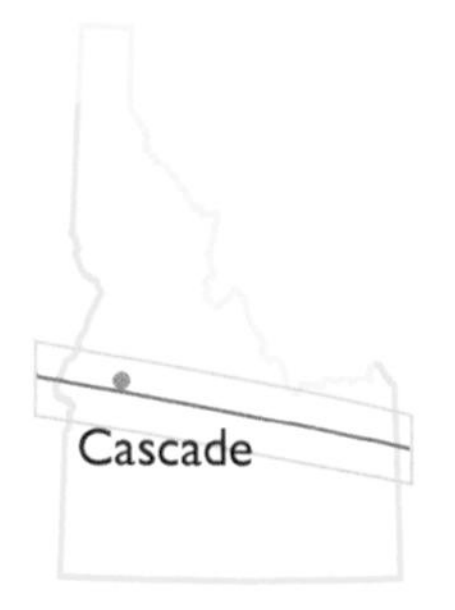

Overview

The high elevations of eastern Idaho have the potential for being excellent total eclipse viewing spots in the summer of 2017. Cascade, a small mountain town located on the shores of Lake Cascade, promises to be one of those spots. Nestled in a valley near the Payette and Boise National Forests, the town has a gas station and restaurants to keep you supplied during the eclipse weekend.

Getting There

Drive north seventy-eight miles from Boise on Highway 55. This is the same route to access Smiths Ferry.

Totality Duration

1 minutes 54 seconds

Notes

Drive around the western side of Lake Cascade for an eclipse-viewing premium location. The cool water of the lake will help reduce atmospheric distortion.

Event	Time (MDT)	Altitude	Azimuth
Sunrise	6:54:00AM	0°	72°
Eclipse Start	10:11:27AM	34°	108°
Totality Start	11:26:52AM	46°	126°
Totality End	11:28:48AM	46°	127°
Eclipse End	12:50:12PM	55°	155°
Sunset	8:38:00PM	0°	287°

Driggs

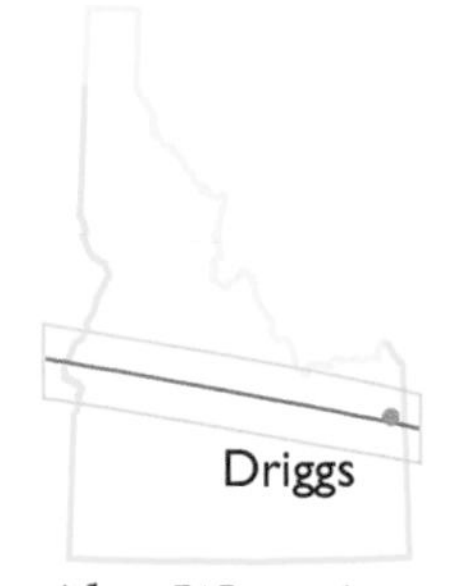

Elevation:	6,109 feet
Population:	1,657 (2013)
Main road/hwy:	Hwy 33

Overview

Driggs is the gateway to the Targhee ski resort near Alta, Wyoming. This vibrant town is enjoyable and one of the secret locations of east Idaho. Busy during the summer, the community is planning multiple total eclipse events. People have been talking about renting out their homes for at least a year once they found out how much homes were being rented for in Jackson, WY. If you take the Swan Valley route, stop at Rainy Creek Country Store for their square ice cream cones at the intersection of US 26 and Highway 31.

Getting There

Driggs can be reached from Idaho Falls (1.5 hours) or Jackson, Wyoming (one hour). Depending on the route you take, you may drive across one or more mountain passes. To avoid the mountain passes, drive from Idaho Falls through Rexburg to Driggs.

Totality Duration

2 minutes 18 seconds

Notes

Visit the Teton Totality website for more information about total eclipse events near Driggs. http://www.tetonvalleyeclipse.com

Event	Time (MDT)	Elevation	Azimuth
Sunrise	6:36:00AM	0°	73°
Eclipse Start	10:16:26AM	38°	113°
Totality Start	11:34:20AM	50°	134°
Totality End	11:36:38AM	50°	135°
Eclipse End	12:59:40PM	58°	167°
Sunset	8:17:00PM	0°	287°

Howe and Berenice

Elevation:	4,829 feet
Population:	220 (2010)
Main road/hwy:	Hwy 33

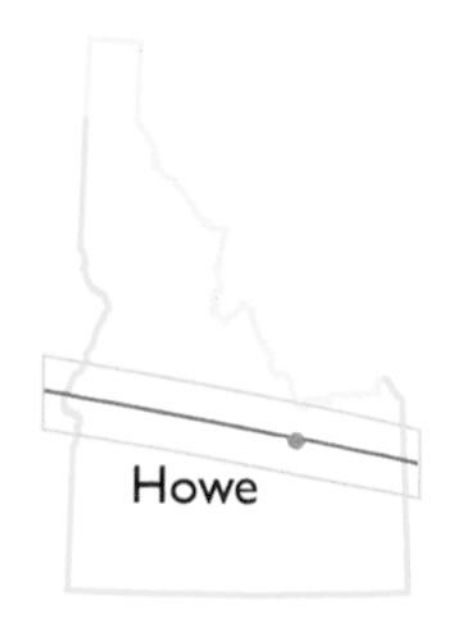

Overview

The views of Saddle Mountain eleven miles north of Howe make a beautiful backdrop for this small Idaho town. Visit the Little Lost Store and the adjacent café to speak with locals about eclipse-viewing access in the area. Highway 33 is small, and it may overload on the day of the eclipse. The community of Howe is a jumping-off point to explore Highway 33. The total eclipse will darken most of this road.

Getting There

Drive sixty miles west from Idaho Falls on US 20/26 to Butte City. Turn right (north) on Highway 33 and drive another twenty miles to Howe. The community of Berenice can be reached by driving approximately five miles northeast along the network of roads in the area.

Totality Duration

2 minutes 11 seconds

Notes

You can either access Howe from Butte City or Terreton.

Event	Time (MDT)	Altitude	Azimuth
Sunrise	6:43:00AM	0°	73°
Eclipse Start	10:14:14AM	37°	111°
Totality Start	11:31:18AM	49°	131°
Totality End	11:33:28AM	49°	131°
Eclipse End	12:56:05PM	57°	162°
Sunset	8:25:00PM	0°	287°

Idaho Falls

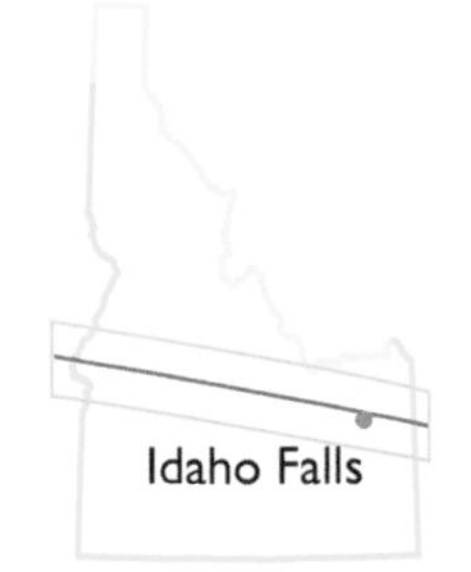

Elevation:	4,705 feet
Population:	58,292 (2013)
Main road/hwy:	I-15

Overview

Idaho Falls is the largest city in eastern Idaho and approximately twenty-three miles south of the total eclipse centerline. Idaho Falls is an excellent starting point to find a location to view the eclipse. Located directly on Interstate 15, this city is one of the major access points to northwest Wyoming. The BBC is planning to broadcast from Idaho Falls. The entire world will focus here in August.

Getting There

Fly or drive into Idaho Falls. Allegiant, Delta, and United all have direct flights to Idaho Falls from major cities. The lowest airfares are found on Allegiant. Use their direct website for flights. Idaho Falls is a three-hour drive from Salt Lake City on I-15.

Totality Duration

1 minutes 49 seconds

Notes

Be aware that the entire eclipse-chaser world is headed to somewhere between Idaho Falls and Casper, Wyoming, to view the eclipse. Do not start late for Idaho Falls from Salt Lake City or Montana. You may find yourself in heavy traffic.

Event	Time (MDT)	Altitude	Azimuth
Sunrise	6:40:00AM	0°	73°
Eclipse Start	10:15:10AM	37°	112°
Totality Start	11:32:59AM	49°	132°
Totality End	11:34:48AM	50°	133°
Eclipse End	12:58:01PM	58°	165°
Sunset	8:21:00PM	0°	287°

Lamont

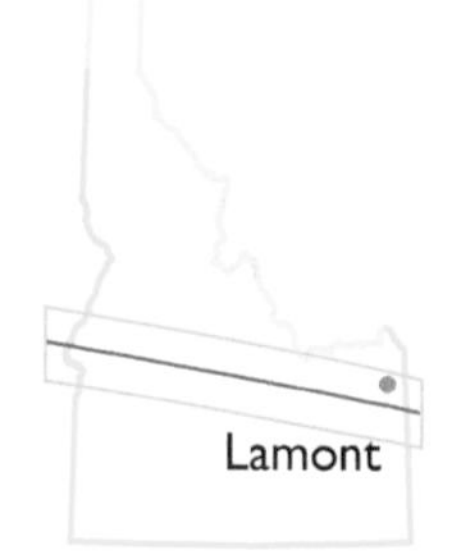

Elevation:	6,030 feet
Population:	Unincorporated
Main road/hwy:	Hwy 33

Overview

There are many locations to view the eclipse directly over the Grand Teton. However, only a few locations are accessible. One of them is in Lamont, Idaho. The Ashton-Tetonia Trail Idaho State Park is near Lamont too. The western side of the Tetons is visible along much of this thirty-mile trail.

https://parksandrecreation.idaho.gov/parks/ashton-tetonia-trail

Getting There

Drive north for ten miles on Highway 33. You will pass through the community of Tetonia. Turn north on Highway 32 and continue north for another 11.5 miles. Turn right (north) on N4700E and then turn right (east) on 700N (Coyote Canyon Road). You can view the eclipse directly over the Grand Teton along this dirt road.

Totality Duration

1 minutes 48 seconds

Notes

The Tetons are twenty-five miles from this location. The drive is easy, as there are many places to pull off the dirt road safely and park. Farming and ranch roads abound in the area.

Event	Time (MDT)	Altitude	Azimuth
Sunrise	6:36:00AM	0°	73°
Eclipse Start	10:16:29AM	38°	113°
Totality Start	11:34:28AM	50°	134°
Totality End	11:36:15AM	50°	135°
Eclipse End	12:59:20PM	57°	167°
Sunset	8:18:00PM	0°	287°

Mackay

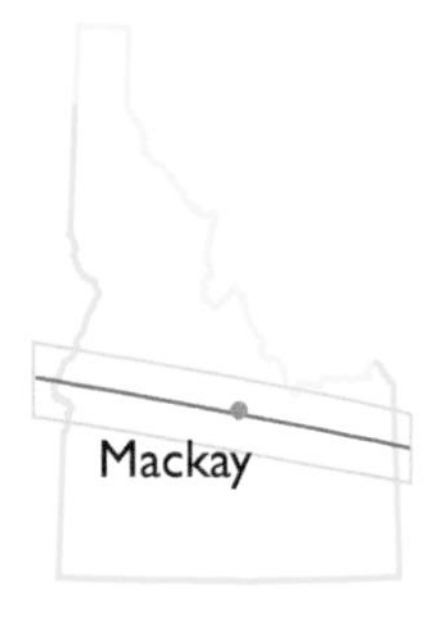

Elevation:	5,906 feet
Population:	506 (2012)
Main road/hwy:	US 93

Overview

Mackay is another Idaho community near the total eclipse centerline. This small hamlet north of Arco is the town closest to Mackay Reservoir. This body of water will be a good place to watch the eclipse from. If possible, watch the eclipse from the western lakeshore. The view over the water will reduce atmospheric distortion, improving the view of the sun's corona.

Getting There

Drive north twenty-six miles from Arco to Mackay along US 93. Mackay Reservoir is 4.5 miles north of the town of Mackay. Access the reservoir by turning into Joe T. Fallini Campground.

Totality Duration

2 minutes 15 seconds

Notes

The Mackay Reservoir is a popular Idaho fishing spot. Visit the Idaho Fish and Game website for more reservoir information:

https://fishandgame.idaho.gov/ifwis/fishingplanner/water/?id=14489

Event	Time (MDT)	Altitude	Azimuth
Sunrise	6:46:00AM	0°	73°
Eclipse Start	10:13:38AM	36°	110°
Totality Start	11:30:19AM	48°	130°
Totality End	11:32:32AM	48°	130°
Eclipse End	12:54:52PM	57°	161°
Sunset	8:27:00PM	0°	287°

Midvale

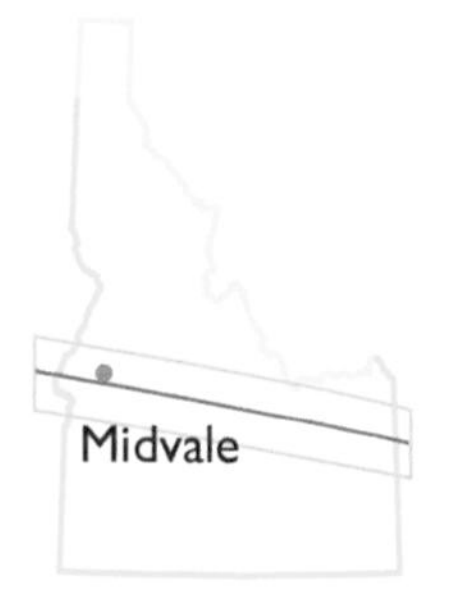

Elevation:	2,543 feet
Population:	160 (2013)
Main road/hwy:	US 93

Overview

Midvale is one of the western communities near the centerline of the total eclipse path in Idaho. Although the official town population is small, the larger community in the zip code 83645 has a population of 636 people according to the town's website:

http://midvaleidaho.com

The abandoned, corrugated steel building near the Country Coffee Cabin may make for an interesting eclipse photograph.

Getting There

Drive north from Boise on I-84 and turn north on US 95 at Fruitland. Drive thirty-eight miles north to reach Midvale.

Totality Duration

2 minutes 4 seconds

Notes

Visit the official Midvale eclipse page for updates on events and lodging at www.midvaleidaho.com/Midvale_Eclipse_Info.html.

Event	Time (MDT)	Azimuth	Elevation
Sunrise	6:57:00AM	0°	72°
Eclipse Start	10:10:42AM	33°	107°
Totality Start	11:25:43AM	45°	125°
Totality End	11:27:50AM	45°	126°
Eclipse End	12:48:55PM	55°	153°
Sunset	8:41:00PM	0°	287°

Mud Lake & Terreton

Elevation:	4,790 feet
Population:	358 (2010)
Main road/hwy:	US 93

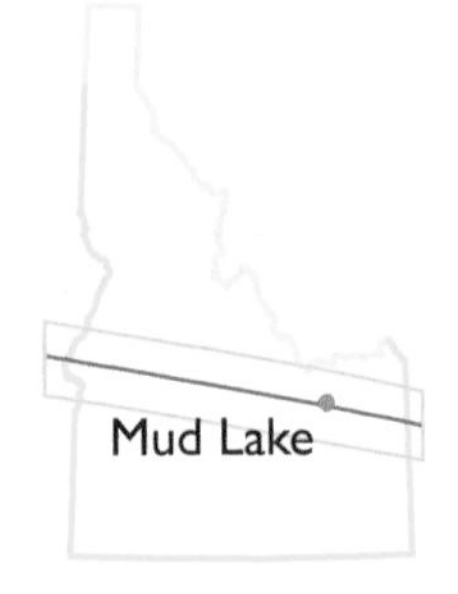

Overview

The communities of Mud Lake and Terreton are along the centerline of the total eclipse in eastern Idaho. Mud Lake is 3.3 miles north of Terreton. During August when temperatures can rise to one hundred degrees, any large body of water in Idaho will be a cooler location to enjoy the total eclipse. Mud Lake has a fairly dry climate and will likely have clear skies.

Getting There

Drive north from Idaho Falls for twenty-four miles on I-15 and turn west on Highway 33. Drive twelve miles to the communities of Mud Lake and Terreton to access Mud Lake.

Totality Duration

2 minutes 18 seconds

Notes

Visit the Jefferson County website for official eclipse information: http://www.co.jefferson.id.us/eclipse.php

Event	Time (MDT)	Altitude	Azimuth
Sunrise	6:41:00AM	0°	73°
Eclipse Start	10:14:55AM	37°	111°
Totality Start	11:32:09AM	49°	132°
Totality End	11:34:27AM	49°	132°
Eclipse End	12:57:06PM	57°	164°
Sunset	8:23:00PM	0°	287°

Rexburg

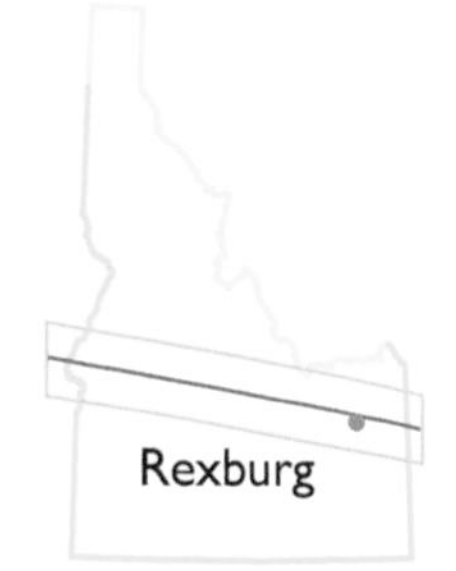

Elevation:	4,865 feet
Population:	26,520 (2013)
Main road/hwy:	US 20

Overview

This small Idaho city is only six miles off the center path of the total eclipse. This makes Rexburg one of major viewing locations in eastern Idaho. The BYU-Idaho campus is hosting planetarium shows, eclipse lectures, kids activities, a star-observing party, and a solar telescope viewing on the day of the eclipse. How popular is Rexburg as an eclipse viewing spot? Rooms at the Super 8 motel were selling for $855 in late March.

Getting There

Rexburg can be reached by driving north twenty-eight miles on US 20.

Totality Duration

2 minutes 17 seconds

Notes

Expect this city to be inundated with people on the weekend of the eclipse. Visit the official Rexburg and BYU-Idaho total eclipse sites for more information: www.rexburgeclipse.com, www.byui.edu/eclipse-2017/events.

Event	Time (MDT)	Altitude	Azimuth
Sunrise	6:38:00AM	0°	73°
Eclipse Start	10:15:41AM	38°	112°
Totality Start	11:33:13AM	49°	133°
Totality End	11:35:30AM	50°	134°
Eclipse End	12:58:19PM	57°	165°
Sunset	8:20:00PM	0°	287°

Rigby

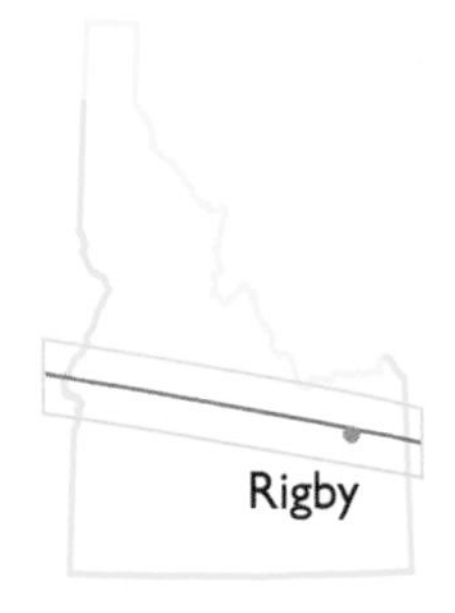

Elevation: 4,856 feet
Population: 3,945 (2010)
Main road/hwy: US 20

Overview

Rigby is famous for being the birthplace of television, originally conceived by Philo Taylor Farnsworth. The town is planning a Main Street event for the total eclipse. You may be able to find campsites in the Bureau of Land Management Idaho Falls District, as all sites are first come, first served. This community and the surrounding cities have been identified by NASA as one of the top four sites in the US to watch the total eclipse.

Getting There

Drive north for fifteen miles from Idaho Falls on US 20 to reach Rigby.

Totality Duration

2 minutes 16 seconds

Notes

Visit the official Rigby eclipse website for more information: http://totalityawesomerigby.com

Event	Time (MDT)	Altitude	Azimuth
Sunrise	6:39:00AM	0°	73°
Eclipse Start	10:15:27AM	38°	112°
Totality Start	11:33:01AM	49°	133°
Totality End	11:35:16AM	50°	133°
Eclipse End	12:58:12PM	57°	165°
Sunset	8:20:00PM	0°	287°

Smiths Ferry

Elevation:	4,554 feet
Population:	75 (2010)
Main road/hwy:	I-15

Smiths
Ferry

Overview

Smiths Ferry is a small hamlet north of Boise along Idaho highway 55, also known as the Payette River Scenic Byway. The community began receiving requests for eclipse reservations years ago due to its unique location. Situated on the Payette River, the town is directly under the centerline of the totality. The two-lane road to this town travels through the mountains.

Getting There

Drive sixty miles north from Boise on Highway 55 to reach the community of Smiths Ferry.

Totality Duration

2 minutes 11 seconds

Notes

Leave early to drive to Smiths Ferry, as in days before the eclipse. Expect that a significant number of people will drive north from Boise (population 214,237) toward Smiths Ferry for the eclipse on August 21. This is one of only three routes north from Boise to view the eclipse.

Event	Time (MDT)	Altitude	Azimuth
Sunrise	6:55:00AM	0°	72°
Eclipse Start	10:11:15AM	34°	108°
Totality Start	11:26:37AM	46°	126°
Totality End	11:28:48AM	46°	127°
Eclipse End	12:50:11PM	55°	155°
Sunset	8:38:00PM	0°	287°

Stanley

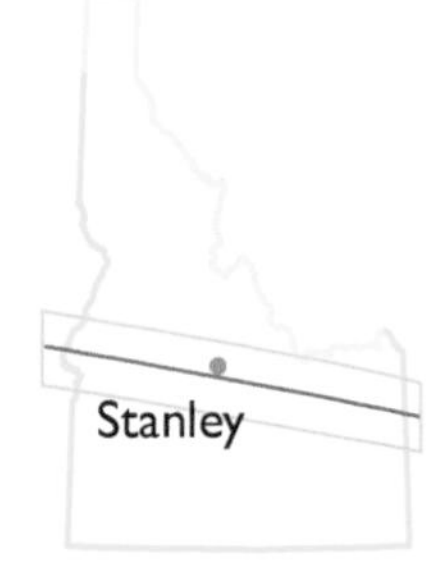

Elevation:	6,253 feet
Population:	69 (2013)
Main road/hwy:	Route 75

Overview

Stanley is another Idaho town situated almost directly under the centerline of the totality. This town is the gathering place for this sparsely populated area of Idaho. The western shore of Redfish Lake will be a good vantage point, as the cool lake water will help reduce atmospheric distortion for eclipse photography. According to the Post Register, Redfish Lake Lodge near Stanley received its first eclipse reservation phone call in 2005. Visit the Mountain Village Mercantile for groceries and other supplies.

www.mountainvillage.com/services/mercantile/

Getting There

Drive sixty-two miles north from Sun Valley on Highway 75 to reach the community of Stanley.

Totality Duration

2 minutes 12 seconds

Notes

Leave early to drive to Stanley. Route 75 is a two-lane highway. Expect heavy traffic early on the morning of the eclipse.

Event	Time (MDT)	Altitude	Azimuth
Sunrise	6:50:00AM	0°	73°
Eclipse Start	10:12:22AM	35°	109°
Totality Start	11:28:18AM	47°	128°
Totality End	11:30:30AM	47°	129°
Eclipse End	12:52:17PM	56°	158°
Sunset	8:33:00PM	0°	287°

Sun Valley & Ketchum

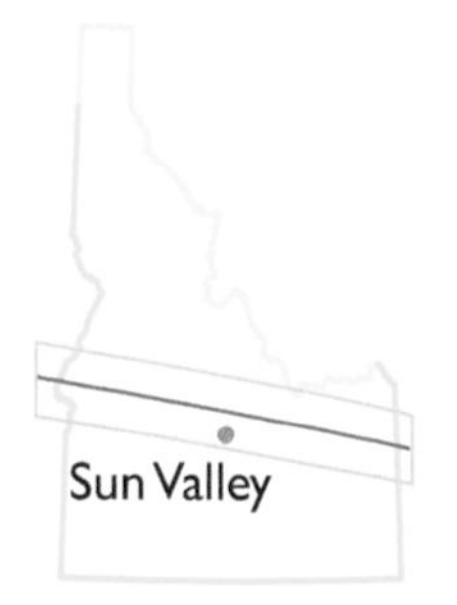

Elevation:	5,945 feet
Population:	1,408 (2013)
Main road/hwy:	Hwy 75

Overview

Sun Valley and Ketchum are one of the most popular and famous of the ski resorts in Idaho. During the summer, outdoor enthusiasts enjoy golf, fishing, and mountain sports. Although these towns are near the edge of the total eclipse path, they are good starting points to access eclipse locations farther north along Highway 75.

Getting There

Drive west from Idaho Falls on US 20 for 120 miles. Turn right (north) on Highway 75 on Gannett Rd. and travel approximately twenty-nine miles to Ketchum. From there, you can access popular destinations in Sun Valley.

Totality Duration

1 minutes 13 seconds

Notes

Visit the official Sun Valley eclipse website for blog updates on the latest events:

http://visitsunvalley.com/events/2017-total-solar-eclipse

Event	Time (MDT)	Altitude	Azimuth
Sunrise	6:49:00AM	0	73°
Eclipse Start	10:12:38AM	35°	109°
Totality Start	11:29:36AM	47°	128°
Totality End	11:30:46AM	48°	129°
Eclipse End	12:53:35PM	57°	159°
Sunset	8:30:00PM	0°	287°

Tetonia

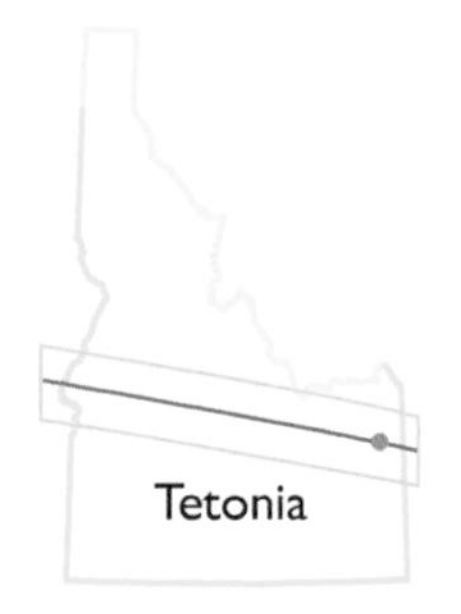

Elevation:	6,047 feet
Population:	266 (2013)
Main road/hwy:	Hwy 33

Overview

Tetonia is a small community in eastern Idaho. It is one of the top viewing spots to watch the total eclipse. The one major advantage Tetonia has over other locations is the eclipse can be viewed over the cathedral group of the Teton Mountains.

Getting There

Drive north 8.5 miles from Driggs on Highway 33 to reach Tetonia.

Totality Duration

2 minutes 10 seconds

Notes

Visit the 2017 Teton Totality group at www.tetontotality.com for updated information about events in and near Tetonia. Finding lodging in Tetonia will be extremely difficult. Most motels and hotels have been sold out for almost a year as of the writing of this guide. Try Airbnb.com or other Internet service for lodging.

Event	Time (MDT)	Altitude	Azimuth
Sunrise	6:36:00AM	0°	73°
Eclipse Start	10:16:28AM	38°	113°
Totality Start	11:34:21AM	50°	134°
Totality End	11:36:34AM	50°	135°
Eclipse End	12:59:35PM	58°	167°
Sunset	8:17:00PM	0°	287°

Weiser

Elevation:	2,129 feet
Population:	5,507 (2010)
Main road/hwy:	US 93

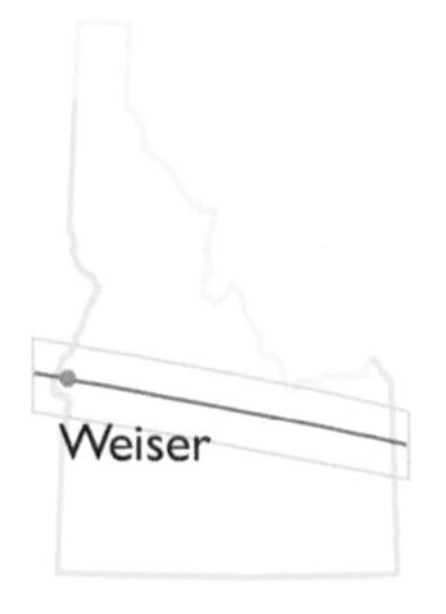

Overview

Located on the Idaho side of the Idaho-Oregon border, the town of Weiser is one of the westernmost communities to enjoy being near the centerline of the total eclipse. The town is planning a series of events starting on August 17 and culminating on August 21. The town has over twenty different events, shows, food competitions, fun runs, fishing tournaments, and even a golf tournament for the total eclipse celebration.

Getting There

Drive north from Boise on I-84 and exit on US 95 in Fruitland. Drive north for nineteen miles to Weiser.

Totality Duration

2 minutes 5 seconds

Notes

Visit the official Weiser total eclipse page for information about their festival:

https://www.weisereclipse2017.com/eclipse-festival/

Event	Time (MDT)	Altitude	Azimuth
Sunrise	6:58:00AM	0°	72°
Eclipse Start	10:10:16AM	33°	107°
Totality Start	11:25:17AM	45°	125°
Totality End	11:27:22AM	45°	125°
Eclipse End	12:48:33PM	55°	152°
Sunset	8:42:00PM	0°	287°

Borah Peak

Elevation:	12,662 feet
Distance:	7.5 miles one way
Main road/hwy:	US 93

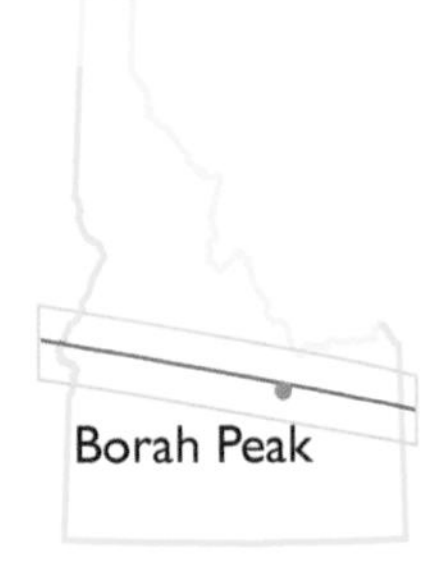

Overview

Located in the Lost River Range, Borah Peak is the highest mountain in Idaho, and it is along the path of the total eclipse. It is also one of the most prominent peaks in the contiguous United States. The most popular route gains over 5,500 feet from trailhead to summit. This summit offers a unique opportunity for highpointers.

Getting There

Drive forty-five miles north from Arco on US 93. Turn right (east) on Birch Springs Road for 3.5 miles to the trailhead parking lot.

Totality Duration

2 minutes 12 seconds

Notes

This is not an easy location to access for families during the total eclipse. It is the most extreme site of any in this guidebook. Do not underestimate the effort to reach the summit. Do not attempt to take children or inexperienced people into the Lost River Range. Depending on the conditions, technical climbing equipment may be necessary. Your safety is your responsibility.

Event	Time (MDT)	Altitude	Azimuth
Sunrise	6:46:00AM	0°	73°
Eclipse Start	10:13:35AM	36°	110°
Totality Start	11:30:07AM	48°	130°
Totality End	11:32:20AM	48°	130°
Eclipse End	12:54:30PM	56°	161°
Sunset	8:28:00PM	0°	287°

Mann Creek State Park

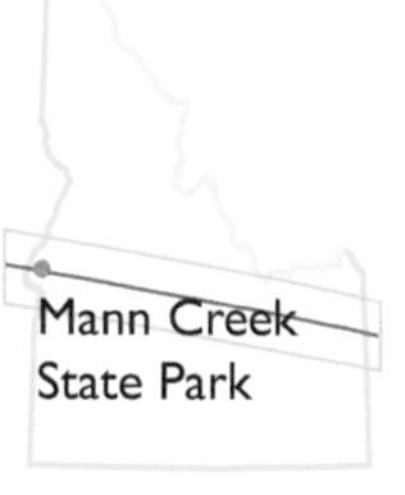

Elevation:	3,300 feet
Distance:	Drive to park
Main road/hwy:	US 95

Overview

Mann Creek State Park at Mann Creek Reservoir is one of the westernmost viewing locations along the centerline of the total eclipse path. This park is at a lower elevation than many other eclipse sites, so it will stay warmer at night than other locations in August.

Getting There

Travel north from Boise along I-84 and take US 95 north to Weiser. From there, travel 10.6 miles to Upper Mann Creek Rd. Turn left (west) and follow the road for 1.5 miles to the campground.

Totality Duration

2 minutes 10 seconds

Notes

Though the park has camping, it has been reserved for some time. Check for campground information at these websites:

https://www.fs.usda.gov/recarea/payette/recarea/?recid=77588
http://idahocampgroundreview.com/mannscreek.html

Event	Time (MDT)	Altitude	Azimuth
Sunrise	6:58:00AM	0°	72°
Eclipse Start	10:10:27AM	33°	107°
Totality Start	11:25:24AM	45°	125°
Totality End	11:27:34AM	45°	126°
Eclipse End	12:48:38PM	55°	153°
Sunset	8:42:00PM	0°	287°

Menan Buttes

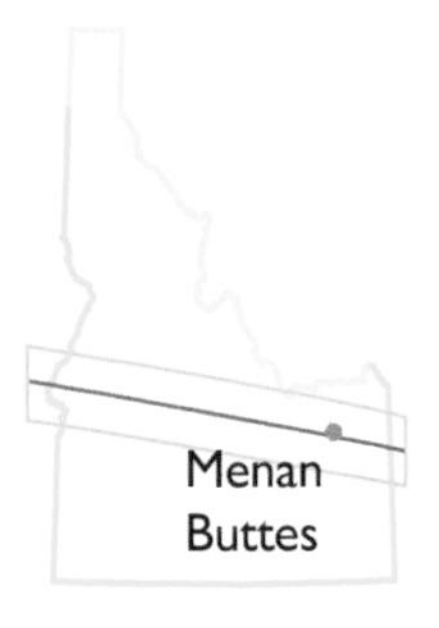

Elevation:	5,619 feet
Distance:	3 miles round trip
Main road/hwy:	Hwy 20

Overview

The Menan Buttes are remnants of volcanoes. These two towering mounds are two of the largest tuff volcanic cones in the world. Designated as a US Natural Landmark in 1980, these buttes rise nearly eight hundred feet above the plains. The steep trail is approximately three miles as a round trip promises to be a unique eclipse spot. Check with local authorities, as access to North Menan Butte may be reserved.

Getting There

Drive west nine miles from Rexburg along Highway 33. Turn left (south) on E. Butte Road. Alternatively, drive to the town of Menan to search for the trailhead that can be difficult for first time-visitors.

Totality Duration

2 minutes 18 seconds

Notes

Be prepared for heat and no shade. Bring plenty of water, and let someone responsible know you are visiting this remote location.

www.top-ten-travel-list.com/blog/exercise/hiking-hotspot-menan-butte-in-rexburg-idaho/

Event	Time (MDT)	Altitude	Azimuth
Sunrise	6:39:00AM	0°	73°
Eclipse Start	10:15:27AM	37°	111°
Totality Start	11:32:55AM	49°	132°
Totality End	11:35:13AM	49°	133°
Eclipse End	12:58:01PM	57°	165°
Sunset	8:21:00PM	0°	287°

Salmon-Challis National Forest

Elevation:	Various
Distance:	Various locations
Main road/hwy:	US 93

Salmon-Challis
National Forest

Overview

The Salmon-Challis National Forest is one of the largest national forests in the lower forty-eight states. The southern section of these 6,618 square miles of forest will be darkened by the total eclipse. The Frank Church—River of No Return Wilderness, the largest wilderness area outside of Alaska, is mostly contained in this national forest too. The intersecting region of the Salmon-Challis, Sawtooth, and Boise National Forests will be covered by the total eclipse.

Getting There

Drive seventy-eight miles north on US 95 to the town of Challis from Arco. From there, access the national forest.

Totality Duration

Eclipse times dependent on location in the forest.

Notes

Refer to local guidebooks for these forests and the wilderness. The Frank Church is aptly named. Do not attempt to travel into this area without proper training and equipment.

Chart times for Challis, ID. Times vary depending on location.

Event	Time (MDT)	Altitude	Azimuth
Sunrise	6:47:00AM	0°	72°
Eclipse Start	10:13:21AM	35°	109°
Totality Start	11:29:59AM	47°	129°
Totality End	11:31:14AM	47°	129°
Eclipse End	12:53:31PM	55°	159°
Sunset	8:31:00PM	0°	287°

St. Anthony Sand Dunes

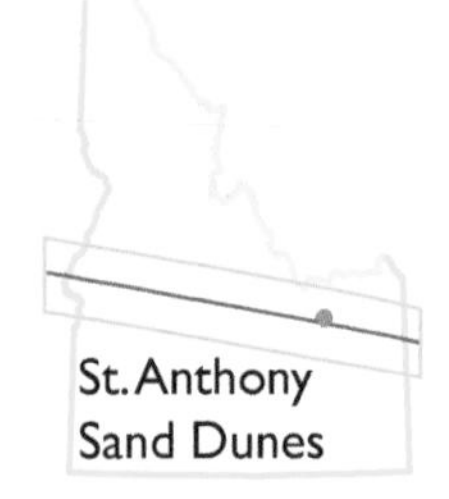

Elevation:	4,967 feet
Distance:	Drive into dunes
Main road/hwy:	US 20

Overview

Located near the town of St. Anthony, this 10,600-acre white quartz sand dunes area is a popular off-road enthusiast recreation area. The tallest sand dune is over four hundred feet tall and is the home of one of the largest herds of wintering elk in the United States. Light from the eclipse on the dunes will be a unique opportunity for photographers.

Getting There

Drive north thirty-nine miles from Idaho Falls on US 20 to reach St. Anthony. Drive through town and turn right (north) on N 1900 E to arrive at St. Anthony Dunes. There is an entrance fee.

Totality Duration

2 minutes 01 seconds

Notes

Check with the BLM Upper Snake Field Office for access and updates: (208) 524-7500 or www.blm.gov/idaho.

Nearby Ashton was rated as another good place for eclipse viewing: http://ashtonidaho.com/events/eastern-idaho-identified-one-best-places-view-august-21-2017s-total-solar-eclipse

Event	Time (MDT)	Altitude	Azimuth
Sunrise	6:38:00AM	0°	72°
Eclipse Start	10:15:47AM	37°	112°
Totality Start	11:33:23AM	49°	133°
Totality End	11:35:24AM	49°	133°
Eclipse End	12:58:14PM	57°	165°
Sunset	8:20:00PM	0°	287°

Remember the Idaho Total Eclipse

August 21, 2017

Who was I with? ____

____

____

____

____

What did I see? ____

____

____

____

____

____

____

____

____

What did I feel? ____

____

____

____

____

____

____

____

What did the people with me think?

Where did I stay?

Enjoy other Sastrugi Press titles

2017 Total Eclipse State Series by Aaron Linsdau

Sastrugi Press has published several state-specific guides to the 2017 total eclipse crossing over the United States. Check the Sastrugi Press website for the various state eclipse books: www.sastrugipress.com/eclipse/

Antarctic Tears by Aaron Linsdau

What would make someone give up a high-paying career to ski alone across Antarctica? This inspirational true story will make readers both cheer and cry. Fighting skin-freezing temperatures, infections, and emotional breakdown, Aaron Linsdau exposes the harsh realities of the world's largest wilderness. Discover what drives someone to the brink of destruction while pursuing a dream.

Adventure One by Aaron Linsdau and Terry Williams, M.D.

What does it take to conceptualize, plan, and enjoy your first expedition? This inspirational book contains hard-won knowledge from both authors about their experiences on expeditions around the world. The information provided in this book is useful whether you plan to climb a high peak, cross a polar plateau, or set out on a never before attempted new trek. (*Available Summer 2017)*

Lost at Windy Corner by Aaron Linsdau

Climbing Denali is a treacherous affair. Avalanches, blinding blizzards, and crevasses have killed experienced teams. What happens when someone decides to climb the mountain solo? In this dramatic story, Aaron describes the choices made and the lessons that were learned as a result. This is more than an adventure story. It teaches defining success on your own terms in business and life. The messages will stay with you long after the end of the book.

Visit Sastrugi Press on the web at www.sastrugipress.com to purchase the above titles in bulk. They are also available from your local bookstore or online retailers in print, e-book, or audiobook form.

Thank you for choosing Sastrugi Press.

"Turn the Page Loose"

About Aaron Linsdau

Aaron Linsdau is a polar explorer and motivational speaker. He energizes audiences with life and business lessons that stick. He delivers a message of courage by building grit and maintaining a positive attitude. Aaron teaches audiences how to eat two sticks of butter a day to achieve their goal. He shares how to build resilience to deal with constant pressure and adrenaline overload.

He holds the world record for the longest expedition in days from Hercules Inlet to the South Pole. Aaron is the second only American to complete the trip alone.

This solo expedition is more difficult than climbing Mount Everest with a team. Being alone dramatically increases the challenge. Aaron uses emotionally stirring stories to show how to overcome obstacles, impossible challenges, and unimaginable conditions. He relates these stories to business challenges and shows how the common person can achieve uncommon results.

Aaron collaborates with organizations to deliver the right message for the audience. He relates his experiences to business realities. Aaron loves inspiring audiences. Book Aaron for your next event today.

"Never Give Up"

Grit • Courage • Attitude • Perseverance • Resilience

Learn more about Aaron Linsdau at:
www.aaronlinsdau.com or www.ncexped.com.

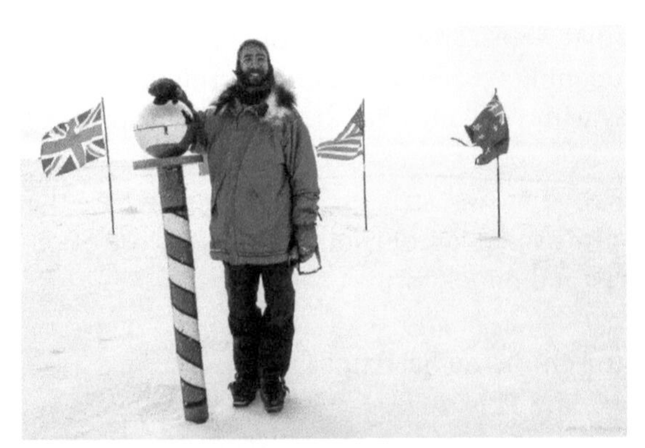

Aaron at the South Pole after 82 days alone in Antarctica.

Smartphone link

CPSIA information can be obtained
at www.ICGtesting.com
Printed in the USA
LVHW01s2115180817
545347LV00010B/322/P

9 781944 986063